D1721528

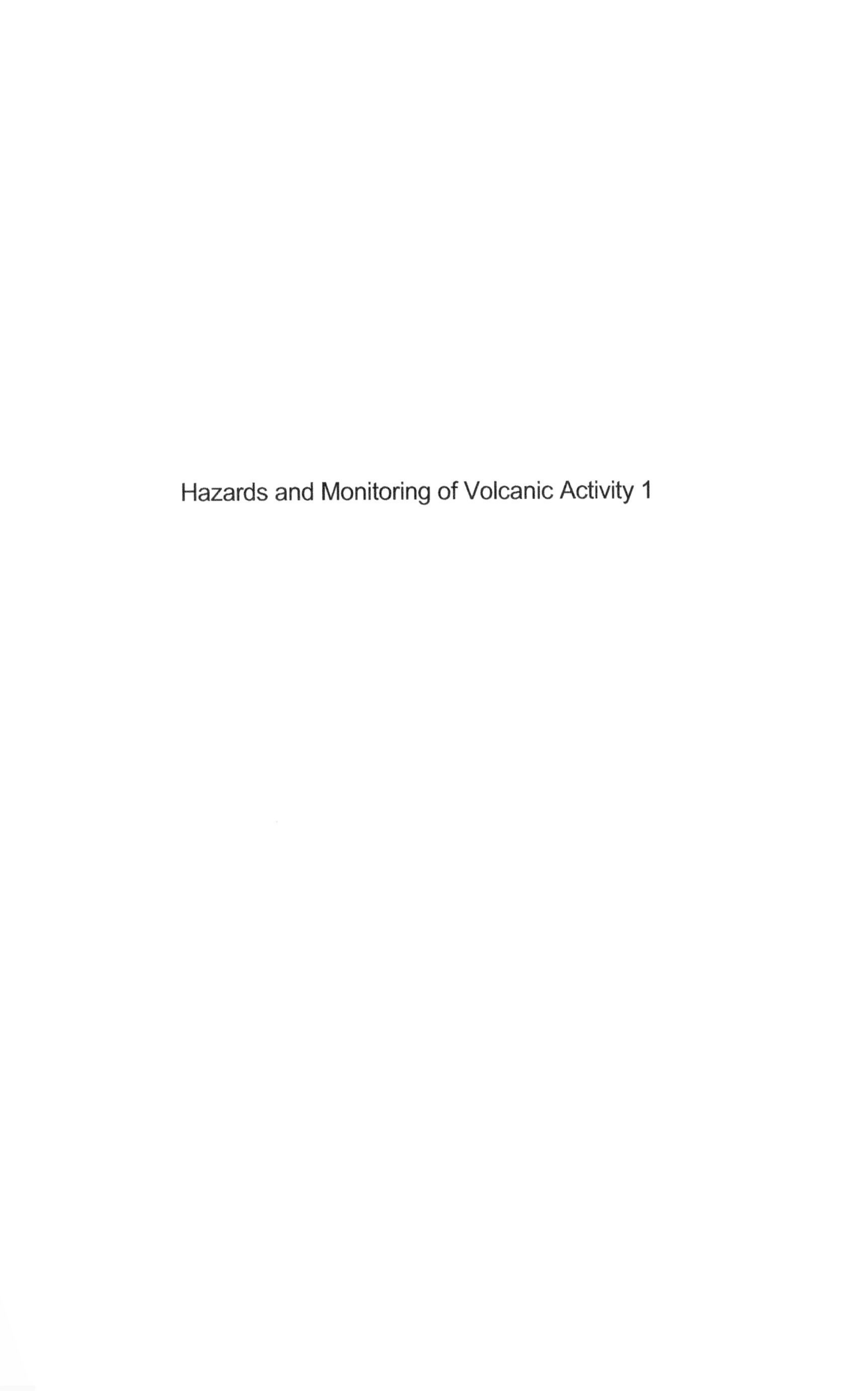

Hazards and Monitoring of Volcanic Activity 1

SCIENCES

Geoscience, Field Director – Yves Lagabrielle

Lithosphere–Asthenosphere Interactions,
Subject Head – René Maury

Hazards and Monitoring of Volcanic Activity 1

Geological and Historic Approaches

Coordinated by
Jean-François Lénat

WILEY

First published 2022 in Great Britain and the United States by ISTE Ltd and John Wiley & Sons, Inc.

Apart from any fair dealing for the purposes of research or private study, or criticism or review, as permitted under the Copyright, Designs and Patents Act 1988, this publication may only be reproduced, stored or transmitted, in any form or by any means, with the prior permission in writing of the publishers, or in the case of reprographic reproduction in accordance with the terms and licenses issued by the CLA. Enquiries concerning reproduction outside these terms should be sent to the publishers at the undermentioned address:

ISTE Ltd
27-37 St George's Road
London SW19 4EU
UK

www.iste.co.uk

John Wiley & Sons, Inc.
111 River Street
Hoboken, NJ 07030
USA

www.wiley.com

© ISTE Ltd 2022
The rights of Jean-François Lénat to be identified as the author of this work have been asserted by him in accordance with the Copyright, Designs and Patents Act 1988.

Any opinions, findings, and conclusions or recommendations expressed in this material are those of the author(s), contributor(s) or editor(s) and do not necessarily reflect the views of ISTE Group.

Library of Congress Control Number: 2022932275

British Library Cataloguing-in-Publication Data
A CIP record for this book is available from the British Library
ISBN 978-1-78945-043-9

ERC code:
PE10 Earth System Science
PE10_5 Geology, tectonics, volcanology
PE10_7 Physics of earth's interior, seismology, volcanology

Contents

Foreword

Claude JAUPART
Institut de physique du globe de Paris, Université de Paris,
Académie des sciences, France

Volcanoes are fascinating because of the beautiful landscapes they form and their superb eruptions. Many books have documented the most catastrophic eruptions, the different eruptive regimes and the physical mechanisms involved. Others have popularized tales of adventurers getting close to volcanic explosions and flows at the risk of their lives. In comparison, the discreet and dedicated work of volcanologists monitoring volcanoes has remained little known. The methods and techniques they use have improved greatly in recent decades and now allow them to predict eruptions with very few limitations. These advances are to be presented and explained in the three volumes of this book.

Volcanoes are built on top of a vast network of underground plumbing, and it is within this network that eruptions occur. Near the surface, the permeable rocks are gorged with water that circulates, heats up at depth and vaporizes under widely varying conditions. The result is a myriad of manifestations, including fumarole fields with changing flows, small earthquakes and ground deformation. This high background noise makes monitoring difficult. An eruption is preceded by the setting in motion of magma from one or more reservoirs located several kilometers underground. This onset is often very discrete and the associated signals are not easily distinguished from the background noise. Once the magma is near the surface, the signs are numerous and leave no room for doubt, but things can move very quickly and it is often too late to evacuate the area. Simply

Hazards and Monitoring of Volcanic Activity 1,
coordinated by Jean-François LÉNAT. © ISTE Ltd 2022.

recognizing that an eruption is imminent is not enough: one must also assess its intensity and regime. Sometimes an eruption can even occur without magma and take the form of phreatic explosions, where the water contained in the surface rocks of the volcano vaporizes explosively. The volcanologist's work does not stop when the eruption begins; they must follow it over time and be able to distinguish between a temporary stop and its true end.

Faced with these multiple challenges, volcanologists have adopted methods that can be divided into two broad categories. The first is the historical study that reconstructs the past eruptions of a volcano and the time intervals between them. Knowing that an eruption covers the deposits of those that preceded it and destroys most of them, establishing a reliable chronology and estimating the volumes ejected rely on particular sampling strategies and frequent round trips between the field and the laboratory. New dating methods had to be developed to determine the ages of deposits older than a few tens of thousands of years. The second category covers all the physical and chemical methods used to determine the deep structure of a volcano and to locate the perturbations that it hosts. Rocks are difficult to penetrate and do not allow us to observe the reservoirs and conduits that feed eruptions. The information we obtain is indirect and often ambiguous. For example, small earthquakes are recorded, but they can be caused by the opening of cracks in a hydrothermal system or by magma that moves or by isolated landslides in places that are not easily accessible. An area of abnormal electrical conductivity can be detected, but it may be weathered rock or rock with water-filled fractures. Volcanologists have improved the uncertainty by combining several methods and have added to their toolbox over the years.

Remarkable progress has been made in the last four decades. Previously, the equipment available was limited to a few heavy and unwieldy devices designed for larger scale studies. Measurements have become much more precise, the number of sensors has increased enormously and the mathematical techniques of analysis have been refined. Well-instrumented observatories have been installed on many active volcanoes and in particular on the three volcanoes of the French national territory in Guadeloupe, Martinique and Reunion Island. Nowadays, a "typical" observatory maintains more than a hundred sensors of all kinds. The last two decades have seen the advent of very efficient satellite tools. However, it should not be inferred that volcanology has become the business of pure measurement experts or laboratory researchers. Knowledge of the special features of each volcano is needed to advance. The active volcano of Santorini in the Cyclades, which grows in the middle of a large caldera, is not monitored in the same way as Piton de la Fournaise on Reunion Island, which rises nearly 7 km above the sea floor and grows on the flanks of the ancient Piton des Neiges. It would be absurd to sample the deposits and install sensors randomly or only in easily accessible locations. Every volcanologist, whether geologist, geophysicist or geochemist, has studied their volcanoes for several years. This patient work has rarely been described. In this book, specialists from all the major disciplines of volcanology share their work and their discoveries. They explain how they decipher and interpret their measurements. One is likely to be surprised by the weakness of the signals detected, which can only be measured with sophisticated instruments, in comparison with the enormity of the eruptive phenomena. But it is thanks to these signals that we are able to travel to the very heart of volcanoes.

Preface

Jean-François LÉNAT
Laboratoire Magmas et Volcans, CNRS, IRD, OPGC, Université Clermont Auvergne, Clermont-Ferrand, France

The impact of natural disasters has become a major concern of our modern societies. Volcanic eruptions, although statistically less deadly and causing less damage than earthquakes or certain atmospheric phenomena, can have devastating local or global effects.

The methods used to determine hazards related to volcanic activity and to monitor the latter is part of many Earth science curricula, both at master and thesis levels.

There are many publications in these areas, but the information is fragmented, requiring teachers to consult a large number of documents to develop their teaching. The aim of this book is to provide them with a single resource, written by specialists, on the methods of monitoring and determining hazards.

The subject is vast, which has led us to present it in three volumes. The first is devoted to geological and historical approaches. The next two are devoted to monitoring methods. The aim of each chapter is not to be encyclopedic. Rather, the intention is to provide the reader with the basic fundamentals of each of the topics covered. On the other hand, each author has taken care to provide bibliographic references that will allow readers to find the detailed information they may need.

This book deals with a scientific field that is constantly evolving. The progress in scientific concepts, approaches, observations and techniques has been spectacular during the last decades. There is no reason why this dynamic should slow down in the future. A logical consequence is that updates should be made periodically to avoid obsolescence of such a book. We therefore hope that it will be useful in the present period and that future editions will enable it to retain its value over time.

March 2022

List of Abbreviations

AI	Ash-Index
AVTIS	All-weather Volcano Topography Imaging Sensor
BLNSS	Base Level Noise Seismic Spectrum
BT	Brightness Temperature
BTD	Brightness Temperature Difference
CAPPI	Constant Altitude PPI
COMET	Center for the Observation and Modeling of Earthquakes, Volcanoes and Tectonics
COSPEC	Correlation Spectroscopy
CRF	Continual Radio-Frequency
DIAL	Differential Absorption Lidar
DOAS	Differential Optical Absorption Spectroscopy
DU	Dobson Unit
EASA	European Aviation Safety Agency
EDM	Electronic Distance Measurement

Hazards and Monitoring of Volcanic Activity 1,
coordinated by Jean-François LÉNAT. © ISTE Ltd 2022.

ENGLN	Earth Networks Global Lightning Network
EUR/NAT	European and North Atlantic
FFM	Failure Forecast Method
FMCW	Frequency Modulated Continuous Wave
FOV	Field of View
FPA	Focal-Plane Array
FTIR	Fourier Transform Infrared Spectroscopy
GEO	Geostationary Earth Orbit
GNSS	Global Navigation Satellite System
GPRI	Gamma Portable Radar Interferometer
GPS	Global Positioning System
IASI	Infrared Atmospheric Sounder Interferometer
IATA	International Air Transport Association
IFOV	Instantaneous FOV
INGV	*Instituto Nazionale di Geofisica e Vulcanologia* (Italian National Institute of Geophysics and Volcanology)
InSAR	Interferometric Synthetic Aperture Radar
LEO	Low Earth Orbit
LF	Linear Fit
LIDAR	Light Detection and Ranging
LP	Long-Period
LPM	Laser Precipitation Monitor

LTA	Long-Term Average
MER	Mass Eruption Rate
MIR	Mid-Infrared
MRR	Micro Rain Radar
MSG	Meteosat Second Generation
NIR	Near Infrared
NOVAC	Network for Observation of Volcanic and Atmospheric Change
NTI	Normalized Thermal Index
OMI	Ozone Monitoring Instrument
OVPF	*Observatoire volcanologique du Piton de la Fournaise* (Volcanological Observatory of Piton de la Fournaise)
PIT	Pixel-Integrated Temperature
PPI	Plan Position Indicator
PPP	Precise Point Positioning
Radar	Radio Detection and Ranging
RSAM	Real-time Seismic Amplitude Measurement
RSEM	Real-time Seismic Energy Measurement
SARA	Seismic Amplitude Ratio Analysis
SEVIRI	Spinning Enhanced Visible and Infrared Imager
SPAC	Spatial Autocorrelation
SSAM	Real-time Seismic Spectral Amplitude Measurement
STA	Short-Term Average

STFT	Short-Term Fourier Transform
TGSD	Total Grain Size Distribution
TIR	Thermal Infrared
TOMS	Total Ozone Mapping Spectrometer
TROPOMI	Tropospheric Monitoring Instrument
ULP	Ultra-Long-Period
USGS	United States Geological Survey
UV	Ultraviolet
VAA	Volcanic Ash Advisories
VAAC	Volcanic Ash Advisory Center
VAG	Volcanic Ash Graphics
VIS	Visible
VLP or VLF	Very-Long-Period or Very-Low-Frequency
VOLDORAD	Volcano Doppler Radar
VT	Volcano-Tectonic earthquakes
WWLLN	World Wide Lightning Location Network

1

Understanding the Geological History of Volcanoes: An Essential Prerequisite to Their Monitoring

Patrick BACHÈLERY
Laboratoire Magmas et Volcans, CNRS, IRD, OPGC,
Université Clermont Auvergne, Clermont-Ferrand, France

1.1. Introduction

How necessary is it to know the past behavior of a volcano in order to monitor it? This chapter will attempt to shed some light on this question. It will present approaches to understanding the eruptive history of a volcano, both in the near and distant past, in order to establish scenarios for future eruptions and a monitoring strategy. Our objective will be to show how improving the knowledge of eruption history, eruption types and timescales is fundamental to improve the prediction and management of eruptions in the short and long term.

For any Earth scientist, it is trite to say that geological time differs from human time. However, this must often be recalled, as we, *Homo sapiens*, have a strong tendency to prefer simple reasoning and to trust a selective perception of the facts supporting our preconceived ideas. In geology, and thus in volcanology, a century that represents the whole of a human lifetime

Hazards and Monitoring of Volcanic Activity 1,
coordinated by Jean-François LÉNAT. © ISTE Ltd 2022.

has little meaning for the complex processes that govern the Earth's evolution, and a long-term process can suddenly change in a catastrophic way. The geological history of volcanoes should help us grasp this.

Today, the monitoring of an active volcano is carried out by the simultaneous use of a set of geophysical and geochemical methods, most often using ground or airborne measurement devices, various samplings and analyses. The deployment of these sensor networks, the acquisition of these data or the realization of these samples must be carried out according to a pre-established strategy based on the knowledge we have of the eruptive past of the volcano, its structure, the assessment of what could be its future activity and, of course, the financial and human resources involved.

Multidisciplinary approaches to tracing magma and fluid transfer from source to surface are now systematic for many volcanoes, whether using natural samples, experiments or the integrated use of geochemical, geophysical and numerical methods. The need for permanent and efficient observation facilities, whether or not they constitute a volcanological observatory (see Chapter 4), has been widely demonstrated. The time is no longer for one-off observations or exploratory missions but for long-term monitoring of volcanoes, both instrumentally and geologically. The implementation of monitoring networks, their spatial configuration, the choice of methods and the choice of parameters to be monitored as a priority are part of a strategy based on the knowledge of the lithology, the structure, and the history of the volcano. Understanding what are, and have been, the processes modifying magmas and their properties – accumulation, storage, transfer, differentiation, crystallization and degassing, as well as the rates and timescales of these processes – is essential to understand the driving forces modulating volcanic activity. A detailed knowledge of the frequency of intrusive and eruptive episodes, their duration and their succession in time according to their nature, is essential to understand the magmatic evolution of a volcano and for a comprehensive assessment of hazards and risks.

The roots of volcanology lie in two letters from Pliny the Younger to the Roman historian Tacitus, describing in detail the eruption of Vesuvius in 79 CE. These observations are now part of world geological history, laying the foundation for the archetype of so-called "Plinian" eruptions. The "modern era" of volcanology probably began with the establishment of the first volcanological observatories at the end of the 19th century in Italy, Japan and then in Hawaii at the beginning of the 20th century (see Chapter 4). The Hawaiian Volcano Observatory (HVO), from its establishment in 1912 by

Thomas A. Jaggar, has of course carried out systematic and continuous monitoring of seismic and ground deformation activity preceding, accompanying and following eruptions, and also a wide variety of other geological, geophysical and geochemical observations and investigations that have greatly improved the understanding of eruptive mechanisms, their causes and their diversity. This understanding, in turn, has been essential to the improvement and diversification of volcano monitoring techniques now routinely used by other volcanological observatories. Geological knowledge and monitoring are intimately linked as are observation, instrumentation, experimentation and modeling. Current developments in new geochemical and petrological approaches, as well as ongoing advances in analytical and imaging techniques, are making it possible to finely document the elemental and isotopic compositions of liquids and minerals, fluid and melt inclusions, as well as textures, with an ever-better precision and spatial resolution. This greatly improves our knowledge of the past magma supply of a volcano and, consequently, of its eruptive past.

The knowledge of the geological history of a volcano is an essential prerequisite for any monitoring of its activity, in order to predict volcanic eruptions or other events. This knowledge is essential for the characterization of hazards (types of eruptions, frequency, cyclicity, changes in eruptive regime and cascade effects) and therefore for the reflection on future eruptive scenarios and the assessment of volcanic risk. It is also essential for the strategic approach to the implementation of volcano monitoring (methods, dimensioning of networks, etc.) and for volcanic risk management.

1.1.1. *Historical volcanology at the crossroads of various disciplines: the example of the Samalas eruption in 1257*

One of the most powerful eruptions in history occurred near the end of the Middle Ages, profoundly affecting the global climate. Its origin, for a long time unknown, was the subject of multiple researches and hypotheses. The source of this "mysterious" eruption was then revealed by the analysis of a set of evidence resulting from a multidisciplinary approach (Lavigne et al. 2013).

The ice core record (see section 1.7.2) showed that one of the largest eruptions recorded in the last 7,000 years was in 1257–1258 (Oppenheimer 2003).

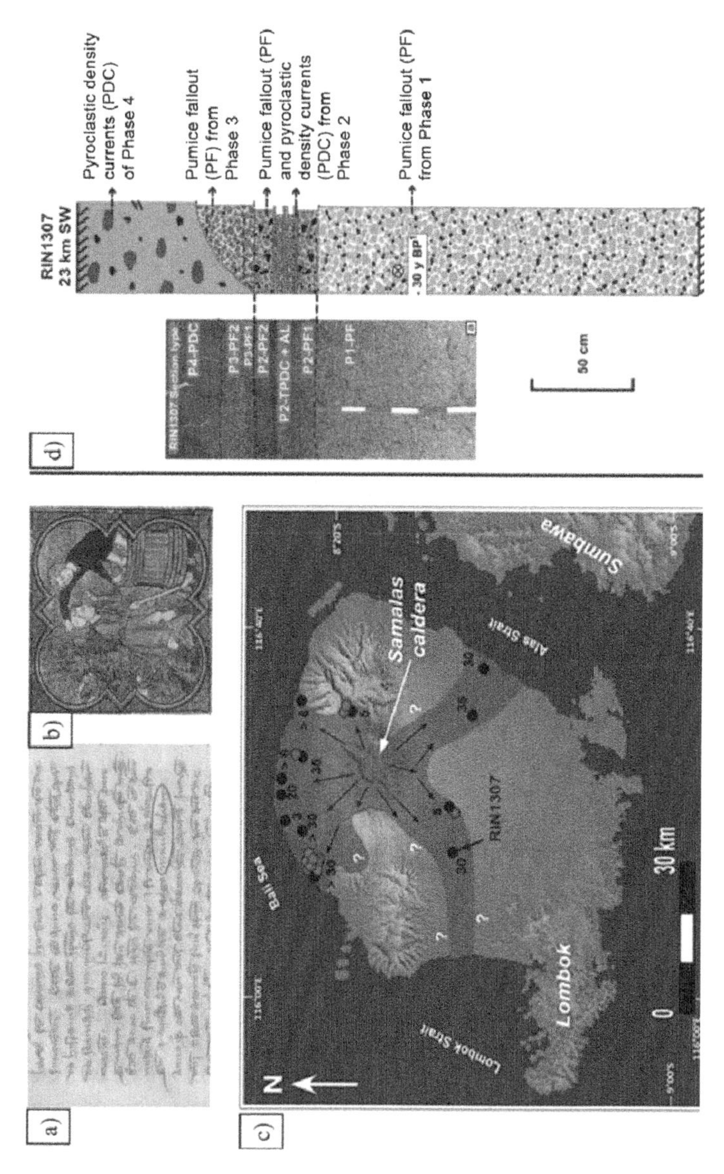

Figure 1.1. *The eruption of the Samalas volcano in Indonesia, in the year 1257, had a major impact on the global climate. The source of this eruption, which remained unknown for a long time, has been found through the interpretation of various types of geological and climatic data and ancient texts. For a color version of this figure, see www.iste.co.uk/lenat/hazards.zip*

COMMENT ON FIGURE 1.1.– *a) and b) Medieval manuscript excerpt and illustration describing the climatic impact observed in 1258 following the eruption of Samalas volcano, and illustration describing the dust veil and climatic disturbances in 1258 (from Guillet et al. 2017). c) Map of Lombok Island (Indonesia) and distribution of pyroclastic flows (purple) attributed to the 1257 eruption around Samalas volcano (from Lavigne et al. 2013) (top right). d) Photograph of deposits from the 1257 eruption and stratigraphic log for the RIN 1307 reference section located 23 km SW of the volcano (from Vidal et al. 2015). See reference articles for details.*

Sulfate concentrations recorded from Greenland and Antarctic ice cores indicate that the eruption generated the largest injection of sulfate aerosols into the stratosphere during our era, far exceeding that of the 1815 Tambora eruption and causing climate disruption (see Chapter 2 for explanation of climate disruption) (Sigl et al. 2015; Vidal et al. 2016; Guillet et al. 2017). The task of identifying the volcano that caused this major eruption took nearly 30 years. The enigma was solved in a convincing way by crossing stratigraphic data (distribution maps of deposits), geomorphological data (3D reconstruction of the ancient collapsed volcano), physical volcanology (volume and flow of the eruption, height of the eruptive column), geochemistry (comparison of the composition of glassy particles collected in the volcano ash with those found in the polar ice and associated with the peak of concentration of sulfate aerosols), ^{14}C dating and the exploitation of a medieval chronicle in ancient Javanese (Babad Lombok). This work identified Samalas, located in the Rinjani volcanic complex on the island of Lombok, Indonesia, as the source of the mysterious eruption (Lavigne et al. 2013). An eruptive column reaching the altitude of 43 km is thought to be responsible for dispersing 33–40 km^3 (DRE – dense rock equivalent) of ash in four eruptive phases including Plinian pumice fallouts and pyroclastic flows, in both hemispheres of the globe, impacting the climate both locally and globally (see Figure 1.1) (Vidal et al. 2015).

This reconstruction of a historical eruption, and its connection to one of the largest Indonesian craters, was made possible by combining different pieces of information that constitute the keys to the reconstruction of the past history of volcanoes: the interpretation of deposits left near or in the region of the volcano that caused their emission, the petrological characterization of these deposits and their comparison with known sources, the use of distant archives, in this case polar ice cores, the recording of past events by dendrochronology, the modeling of climatic data, the use of historical archives and image analysis. The above example, applied to a specific eruption, describes the more global approach that must be taken when reconstructing the geological history of a volcanic edifice as a whole in order to characterize potential future hazards and risks.

1.1.2. *Hazard characterization, geological analysis and future eruptive scenarios*

Knowledge of the geological history of volcanoes forms the basis of current methodology for volcanic hazard assessment. The geological characterization of past eruptions of a volcano, the interpretation of its geological archives and their translation into hazard maps or even volcanic risk maps (see Chapter 3), and their processing by probabilistic approaches are part of the long-term assessment of the activity of a volcano. It is intended to plan strategies, procedures and measurements for monitoring active or potentially active volcanoes, just not to assess the day-to-day status of the volcano (Crandell et al. 1984). These issues enable the necessary dialog between scientists and civil protection authorities and provide essential information to the populations living near volcanoes. The risk assessment is based on the documented history of the volcano under study.

However, understanding the geological history of volcanoes is not completely absent from the day-to-day risk assessment, in addition to the continuous observations and measurements carried out by the geophysical and geochemical monitoring networks. This knowledge of the volcano's past behavior must be considered in the assessment of the daily state of the volcano and its day-to-day evolution, as well as the sequences of eruptive events for similar volcanoes, which can provide information on a range of potential behaviors (Jousset et al. 2012). These insights define the "experience" of the observatory, which still represents the basis for assessing the state of a volcanic system today, even if the development of so-called "expert" computer systems tends to grow.

1.1.3. *Mount St. Helens, May 18, 1980*

The eruption of Mount St. Helens volcano that began on May 18, 1980, made a profound impact on the scientific community working in volcanology. It allowed the first observation and characterization of a previously unknown type of explosive eruption, and it demonstrated that several types of dangerous phenomena can be combined in a very short period of time during a single eruption. Within a few hours, this eruption produced a flank collapse, a direct lateral blast, avalanches and debris flows, pyroclastic flows and large tephra falls. Above all, it was the first highly explosive eruption to benefit from prior knowledge of the volcano's history and from enhanced instrumental monitoring.

The work of Crandell and Mullineaux (1978) is a wonderful illustration of the contribution of research on volcanoes to volcanic risk assessment. Two years before the eruption, Dwight Crandell and Donal Mullineaux published a report in the Geological Survey Bulletin, presenting the results of several years of geological studies of past eruptions of Mount St. Helens (Crandell and Mullineaux 1978). In this work, the nature of past eruptive events, their frequency and the threats they could pose to people and property in case of a future eruption are outlined. This study is part of a larger investigation of the volcanoes of the Cascade Range in the western United States of America, initiated in 1967 by the USGS. It appeared as early as 1975 (Crandell et al. 1975) that Mount St. Helens is the youngest and most active of the Cascade volcanoes. Based on the interpretation of eruption sequences over the last 4,500 years, Crandell and Mullineaux (1978) predicted an eruption of Mount St. Helens "within the next hundred years, and perhaps even before the end of the century". Their prediction proved to be correct, but that is probably not the most important thing. The major contribution of their study lies in the description given of the phenomena associated with a future eruption, and their representation in a volcanic hazard map. Their assessment of volcanic hazards, deduced from the study of deposits left by past eruptions, was an essential contribution to the management of the 1980 eruption and the minimization of the risks associated with this eruption. Based on the 1978 volcanic hazard mapping proposed by Crandell and Mullineaux, an evacuation zone was established by the authorities, saving many lives during the sudden explosion of May 18, 1980.

The first signs of activity at Mount St. Helens appeared on March 16, 1980, with shallow seismic activity located beneath the volcano (Lipman and Mullineaux 1981). In the following days, this seismic activity proved unusual and a monitoring network was installed. On the basis of the volcanic hazard mapping proposed in 1978 by Crandell and Mullineaux (1978), access to the volcano and to potentially exposed areas was limited. A few hundred inhabitants were evacuated. On March 27, the first phreatic explosions were observed at the summit, as well as the opening of numerous fractures associated with the formation of a crater and the initiation of a bulge on the upper part of the northern flank of the volcano (see Figure 1.2).

An intensive program of geophysical monitoring of these events was then set up by USGS researchers and technical staff. The analysis of volcanic risks was further developed on the basis of Crandell and Mullineaux's work. Until mid-May, the high rate of seismicity, the spectacular deformation of the northern flank of Mount St. Helens and the intermittent phreatic eruptions led them to consider the possibility of a major eruption in the near future. These conditions also justified maintaining the restricted area established at the end of March. The catastrophic explosion of May 18 at 8:32 a.m. began without any further warning. Fifty-seven people were killed, mainly by asphyxiation or burns, including scientists and photographers working on the edge of the exclusion zone.

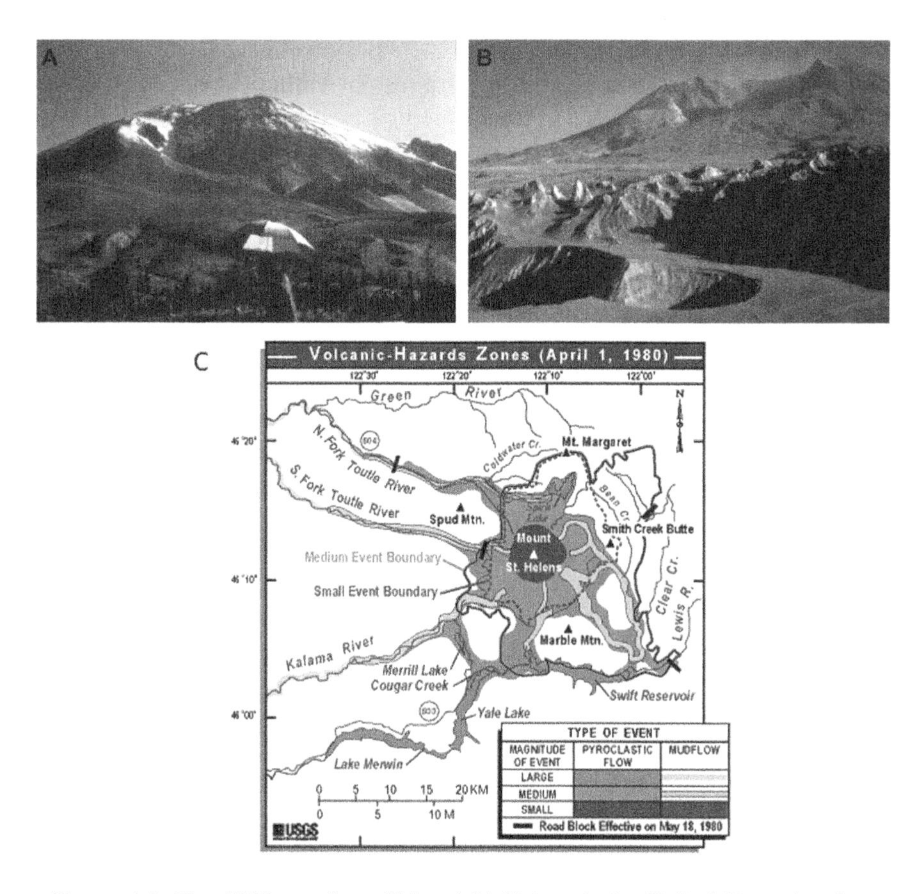

Figure 1.2. *The 1980 eruption of Mount St. Helens in the United States is a key example of how knowledge of the geological history of a volcano can be used to monitor it. For a color version of this figure, see www.iste.co.uk/lenat/hazards.zip*

COMMENT ON FIGURE 1.2.– *a) Bulging of the north flank of Mount St. Helens before the eruption, April 27, 1980 (Cascades Volcano Observatory Photo Archives). b) Mount St. Helens after the May 18, 1980 eruption (September 10, 1980). The volcano has lost about 400 m of its height (USGS Volcano Hazards Program). c) Volcanic hazard zonation for a possible eruption of Mount St. Helens, based on the map proposed by Crandell and Mullineaux (1978), revised to April 1, 1980.*

1.1.4. *Lessons learned from the eruption of Mount St. Helens*

The validity and usefulness of the volcanic hazard assessment and zoning maps, which had been published 2 years earlier, were thus fully demonstrated, albeit with a few important differences. The hazard assessment was based on eruptive events of the past 4,500 years. It assumed that future events from the volcano would be similar in type and intensity to those of that period. The prodigious collapse of the northern flank of the volcano and the resulting debris avalanche and lateral blast had not occurred during the period studied and, even if such events had occurred in the late Pleistocene, they were not known, and therefore not anticipated, in the risk assessment published in 1978. Such a slide of the northern flank of the volcano had, however, been envisaged in the month before the eruption, because of the continuous and significant deformation measured on this flank, but the extent of the debris avalanche and the magnitude of the directed blast were underestimated, affecting a much larger area than originally anticipated (see Figure 1.3).

Among the lessons learned after this eruption, we note the complementarity, as demonstrated here, between geological data and instrumental monitoring data, for the prediction of events and the evaluation of the possible course of the eruption. It will also be noted that, even if one cannot rely exclusively on knowledge of past history, it must be considered as a starting point for a probabilistic approach, for which it is essential to take into consideration possible analogs elsewhere in the world (Lipman and Mullineaux 1981).

Regardless of the volcano, destructive events may exceed those of the past. However, should the largest known eruption of any similar volcano in the world be taken as a reference for establishing a monitoring strategy? Since the 1980 eruption of Mount St. Helens, other work on the recent geologic history of this volcano has complemented Crandell and Mullineaux's study. For example, Pallister et al. (1992) reassessed the eruptive activity during the last 500 years, highlighting variability in magma composition and discussing possible links between the different phases of activity (five major explosive eruptions, including the one in May 1980, and other more moderate ones) and deep inputs of new magma. How does magma chemistry, a major parameter determining the eruptive regime, evolve from one eruption to the next? The knowledge of the functioning of the magma system of the volcano, whatever it is, is a key parameter of the monitoring. At Mount St. Helens, the absence of deep geophysical precursors and the lack of evidence of deep magma reinjection led to the conclusion that the May 1980 eruption resulted from the partial crystallization of a surface reservoir in which the magma had remained for a few centuries to millennia, and thus from the effect of the so-called "second boiling" (supersaturation in CO_2 and H_2O induced by this partial crystallization). The magma reservoir(s) of a volcano is/are formed in a place (rheological and/or density contrast) where the accumulation of magmas, formed by a slow and continuous process at greater depth, is possible. The episodic release of these magmas toward the surface to feed an eruption is a complex process that we cannot fully predict, but it allows, for a single volcanic edifice, to feed emissions that vary in terms of volume and eruptive regime. These variations in eruptive regimes are particularly difficult to predict. We know that they result from changes in the physicochemical properties of magmas, or from external parameters such as the presence of water, the geometry of eruptive conduits or the morphology of the edifice (Cassidy et al. 2018).

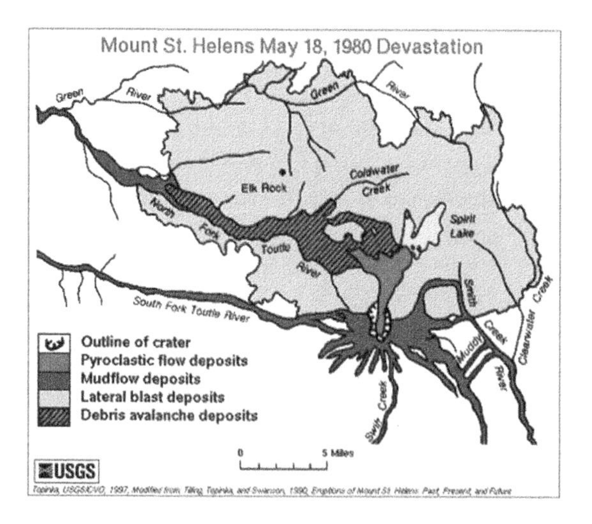

Figure 1.3. *Mapping of deposits from the 1980 eruption at Mount St. Helens (note the difference in scale with map 1.2C) (Topinka, USGS/CVO 1997 adopted from Tilling et al. 1990). For a color version of this figure, see www.iste.co.uk/lenat/hazards.zip*

Figure 1.4. *Two contrasting eruptive styles at Kilauea in Hawaii. a) The lava lake at Halema'uma'u crater in 1893. The lava lake is contained by levees formed by small overflows or weak projections, originating from the lava lake itself (https://www.usgs.gov/media/images/k-lauea-visit-was-a-prelude-revolution; photo Brother Bertram). b) The phreatic eruption of Halema'uma'u on May 18, 1924 (https://www.usgs.gov/media/images/explosive-eruption-column-halema-uma-u-1115-am-may-18-1924; photo Kenichi Maehara)*

1.1.5. *The diversity of eruptive regimes*

If the eruption of Vesuvius in 79 CE has a special place in the history of volcanology, it obviously does not alone account for the reality of the eruptive activity of this volcano. Although it is essential to take into account major paroxysms in order to assess the volcanic risk and to prepare the measures to be implemented for the protection of the population, the forecasting of volcanic activity by the Vesuvius Observatory must take into account all the possible events, including the more common phases of activity, and therefore consider the diversity of eruptive regimes.

The eruption of Mount St. Helens has shown how difficult it is to anticipate changes in the eruptive regime of a volcano. From a purely phreatic eruption, which does not present any significant danger, the eruption can rapidly evolve toward a catastrophic outcome with a major magmatic paroxysm. This suddenness has been demonstrated by many examples beyond Mount St. Helens, for example, Soufrière Hills on Montserrat in 1995 (Kokelaar 2002; Sparks et al. 2002; Voight et al. 2002). This implies a good knowledge of past eruptions, their succession, and also of the diversity of magmas and eruptive regimes within a single eruption.

The variability of eruptive regimes largely results from the diversity of magmas and the conditions of their transfer to the surface. Magmas undergo profound changes in their physical properties due to variations in pressure and temperature conditions during their evolution in the magma reservoir, during their ascent to the surface and during the eruption itself (Sparks 2003). Degassing, cooling and crystallization during magma ascent induce a drastic change in their physical properties, especially viscosity. Degassing conditions strongly determine the regime under which magma rises and will erupt to the surface (Moretti et al. 2018). Active magma systems interact strongly with their environment, inducing deformation and rupture of the bedrock and disturbances of subsurface aqueous systems. These various processes and interactions that have direct physical and phenomenological effects during eruption must be identified from petrological and geochemical studies of previous eruptions. They serve as a basis for modeling to improve the understanding of the physics of the processes involved. Given the uncertainties and complexity of nonlinear systems such as volcanoes, it is still very difficult to make accurate predictions (Sparks 2003). For eruptions, predictions and hazards must be expressed in probabilistic terms and take into account uncertainties. This has important consequences in the management of volcanic crises, which must integrate the human parameter, whose importance is great.

This implies that the scientists in charge of crisis management must have an appropriate dialog with the authorities and the public and must be prepared for a probabilistic approach.

What scenarios are or should be considered? This question inevitably arises when a volcano enters a period of instability. For volcanological observatories or scientists in charge of monitoring, it is then essential to be able to base their reflections and monitoring strategies on a strong geological knowledge allowing for anticipation of a possible scenario for the coming eruption. The 1976 La Soufrière crisis in Guadeloupe clearly demonstrated the scientific and communication biases induced by this lack of knowledge. The scientists did not have a sufficient geological framework to interpret the signals. An analysis of this crisis has been made a posteriori, in the light of our knowledge and with current tools, in particular those of Bayesian statistical analysis. This analysis shows that if, at the climax of the crisis, the probability of a magmatic intrusion was high, the prediction of the outcome of the eruption remained uncertain (Hincks et al. 2014; Komorowski et al. 2018).

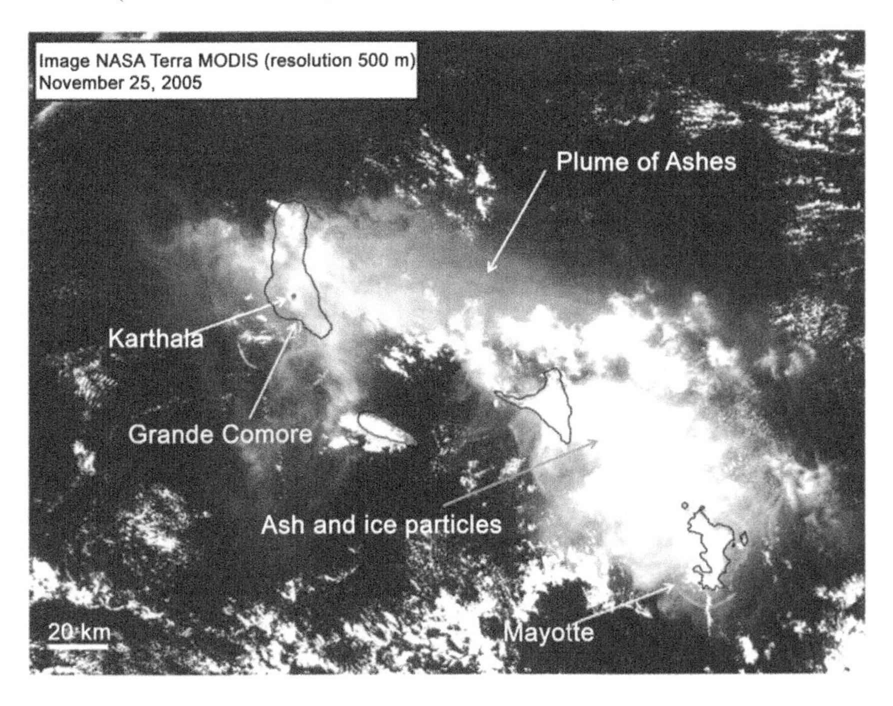

Figure 1.5. *NASA Terra MODIS image of the ash plume from the violent strombolian eruption at Karthala volcano, Grande Comore, on November 25, 2005. For a color version of this figure, see www.iste.co.uk/lenat/hazards.zip*

COMMENT ON FIGURE 1.5.– *The Comoros archipelago (in the Indian Ocean) is composed of the islands of Mayotte, Anjouan, Moheli and Grande Comore. Their outline is shown in black. The red dot marks the position of the Karthala crater, origin of the eruption. The 2005 eruption marks a significant change in the eruptive regime of this volcano, which usually has mostly effusive eruptions. The ash plume extended for about 280 km and reached an altitude of ~12 km (Bulletin of the Global Volcanism Network, vol. 30, no. 11 - November 2005).*

Variability in eruptive regimes is not limited to volcanoes with relatively low eruption frequency. Volcanoes with frequent activity, such as Etna (Behncke et al. 2008; Calvari et al. 2018; La Spina et al. 2019), or even quasi-continuous activity such as Stromboli (Calabrò et al. 2020) or basaltic shield volcanoes such as Kilauea in Hawaii or Piton de la Fournaise on Reunion Island, also experience multiple eruptive regimes, often during a single eruption. For the latter, lava lake activity, powerful lava fountains, violent strombolian explosions or phreatic or phreatomagmatic activity may alternate, whereas these volcanoes are often considered essentially effusive (see Figure 1.4). Deposits of phreatic and phreatomagmatic explosions such as Keanakāko'i ash at Kilauea (Swanson et al. 2012, 2014) or Cendres de Bellecombe (Bellecombe Tephra sequence) at Piton de la Fournaise (Bachèlery 1981; Ort et al. 2016) are probably the best examples of this eruptive variability, as are the violent, phreatomagmatic strombolian eruptions of Karthala in 2005 in Grande Comore (Bachèlery et al. 2016) (see Figure 1.5).

1.2. Relative and absolute dating and the importance of timescales: chronology of eruptions

The aim here is not to describe the dating methods but to recall their importance in determining the main periods of building of the volcano as well as in understanding the succession in time of recent eruptive episodes, and to show what are the possible methods based on a few selected examples.

The time factor is a key element in the behavior of a volcano. As we have already stated, a volcano irregularly ejects magmas formed by slow and deep processes acting continuously. This transition from a continuous process to an episodic and irregular process is determined by the transit of magmas through the mantle, the crust and the volcanic edifice, and thus by their storage and physico-chemical evolution within one or more magma reservoirs. Time plays a major role in these processes, hence the importance of understanding the past activity of a volcano. In an eruptive sequence, it is essential to be able to determine the frequency of eruptions. This frequency makes it possible to assess how quickly the batch of magma that can feed the next eruption can be reconstituted, its degree of chemical evolution, and therefore the rate of recurrence of the different types of eruptions. These data allow us to reconstruct past eruptive cycles, to know their duration and the sequence of events within the same cycle. From this knowledge, we can determine the state of the volcano and what we can expect from its activity in the near future. This is essential data for volcano monitoring.

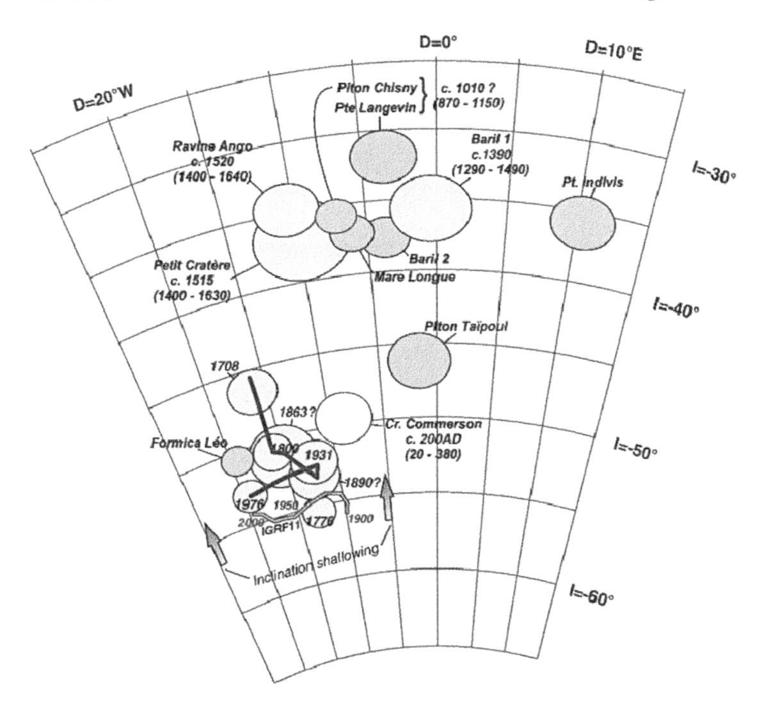

Figure 1.6. *Archeomagnetic directions obtained for recent lava flows from the Piton de la Fournaise volcano (on Reunion). The secular variation of Earth's magnetic field is used here to date recent lava flows[1] (according to Tanguy et al. 2011). For a color version of this figure, see www.iste.co.uk/lenat/hazards.zip*

1 Dating by archeomagnetism: volcanic rocks record the ambient magnetic field during their cooling. The secular variation of Earth's magnetic field is used to date recent lava flows (see

Reconstructing this history implies having a set of data that allow us to assess the time within an eruptive sequence. This requires an effort of systematic dating of events, either by absolute dating methods that allow us to accurately locate the events in time, or by relative dating that allows us to locate the events between them without absolute reference. Several methods allow the absolute dating of volcanic rocks, their choice depending on the age range targeted and the nature of the material to be dated.

There is a wide range of radiometric methods for dating the rock itself or the minerals or glass within a tephra. The most frequently used are the $^{40}Ar/^{39}Ar$, K-Ar and Rb-Sr isotopic methods, the methods using the isotopes of the decay chain of Uranium and Thorium, the exposure to cosmogenic nuclides and the fission traces. The ^{14}C method is most frequently used for "young" ages (<50 ka), but it requires finding fragments of carbonized wood from trees burned by lava flows (Rubin 1987), or shell or coral fragments, especially for tephra sampled in marine environments (Köng et al. 2016).

The relative dating of events can be tackled with the help of the stratigraphy of pyroclastic and epiclastic flows or deposits, the biostratigraphy, the stable isotopes ($\delta^{18}O$), the geographical or geomorphological relationships between two events, the archeomagnetism and paleomagnetism (see Figure 1.6), the geomorphology of volcanic cones or lava flow surfaces, or even, in favorable conditions, from the vegetation cover (see Box 1.1). Other methods are more rarely used in volcanology, such as thermoluminescence, hydration crust development, ice cores, varves, dendrochronology or lichenometry.

Figure 1.6). The brown line connects historically dated flows (95% confidence circles in yellow). The beige circles represent lava flows that are historically undated or whose proposed date is questionable. The light yellow circle represents the tephra emitted by the Commerson crater (dated by ^{14}C). The secular variation on Reunion appears significantly different from that observed in Europe. In particular, the low amplitude of the directional variations observed over the last 250 years (inclination (I) within the limits of –50° and –55°, declination (D) from –13° to – 9°) makes any dating archeomagnetic difficult for this period. In contrast, a larger directional variation exists for the period 1750–1000 CE. This approach makes it possible to determine the date of recent flows but not precisely.

The colonization of new lava flows emitted by volcanoes in the intertropical zone is the focus of much attention, both in the emerged domain (Ah-Peng et al. 2007; Albert et al. 2020) and in the marine domain (Zubia et al. 2018). On land, rapid plant growth in these warm and humid contexts allows for rapid invasion of new flows by lichens, mosses and shrubs. This is used for dating purposes for recent lava flows with unknown ages. Dating methods such as lichenometry or dendrochronology can be used considering that, in these biologically favorable contexts, the development of vegetation on the flows (see Figure 1.7) depends mainly on the age of emplacement of the lava flow (Atkinson 1971).

Figure 1.7. *Differences in vegetation coverage of recent flows on the eastern flank of Piton de la Fournaise. The most recent flows are largely devoid of vegetation and appear dark; the oldest (a few hundred years) are entirely covered*

These approaches have been developed on Reunion Island where Piton de la Fournaise is frequently active with several eruptions per year (Lénat and Bachèlery 1988; Peltier et al. 2009a). Although the frequency of eruptions is well known for the last decades, mainly since the establishment of a volcanological observatory in 1980, this is not the case for the historical eruptions that took place between the 18th and mid-20th centuries. The "biological" dating is therefore one of the tools used to enrich the knowledge of the eruptive history of this volcano. For example, Albert et al. (2020) use the strong correlation between the maximum diameter at the base of *Agarista salicifolia* and the age of lava flows. This relationship, established from flows of known age (historical observation or ^{14}C dating), can be used to estimate the age of recent undated flows (see Figure 1.8).

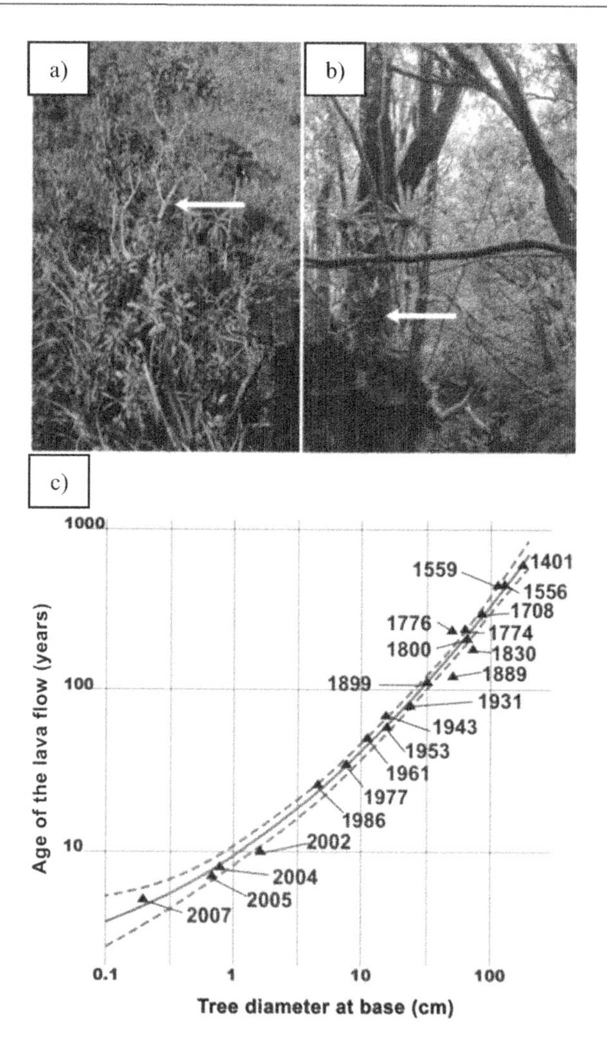

Figure 1.8. *The different growth stages of Agarista salicifolia (marked by white arrows) are used to date recent flows from the Piton de la Fournaise volcano*

COMMENT ON FIGURE 1.8.– *a) A young agarista installed on a 1986 lava flow. b) Agarista installed on a 1708 lava flow. c) Calibration curve for dating volcanic activity based on the relationship between lava flow age and maximum diameter at the base of Agarista salicifolia (according to Albert et al. 2020).*

Box 1.1. *Dating of recent lava flows at Piton de la Fournaise on Reunion Island*

1.3. Frequency of eruptions, eruptive cycles and future eruption scenarios

The eruptions in one volcano can sometimes have very different characteristics. The chemical composition of the magma and its evolution over time are, of course, key parameters, but they are not the only ones. The morphological and structural evolutions of the edifice also determine the functioning of the volcano. Therefore, if we want to evaluate what the eruptive activity of a volcano could be, it is not possible to refer only to the most recent period. Santorini, in the Aegean Sea, or Italy's Vesuvius are good examples.

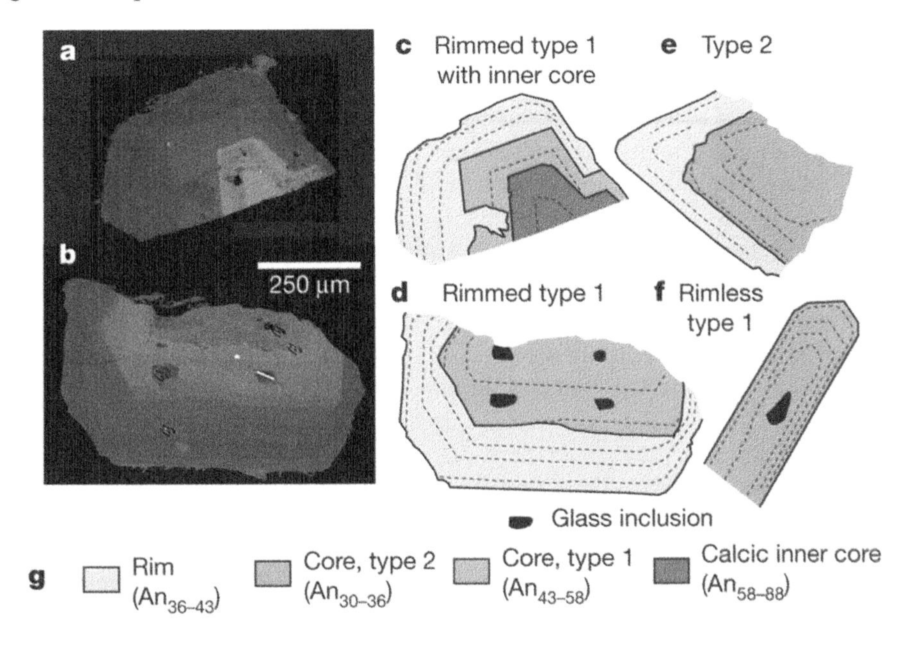

Figure 1.9. *Crystals in lavas and pyroclastics are archives of the evolution of magmas within volcanoes, as shown here by the compositions of plagioclase crystals in the pumice from the Minoan eruption in Santorini. For a color version of this figure, see www.iste.co.uk/lenat/hazards.zip*

COMMENT ON FIGURE 1.9.– *Chemical variations measured on zoned plagioclase crystals from the Minoan eruption at Santorini (~1600 BCE) have shown that, despite 18,000 years separating the Minoan eruption to the previous major eruption, most Minoan magma crystals record the recharging of the magma reservoir by large volumes of silicic (and some mafic) magma less than 100 years before the eruption and that mixing between the different magmas took place a few months before the eruption was initiated. These are essential pieces of data for monitoring Santorini-type volcanoes (according to Druitt et al. 2012).*

The reconstruction of long time series of eruptions allows to evaluate the regularity over time of the behavior of these volcanoes, and in particular the existence of cycles including different types of eruptions and eruptive regimes. It also allows us to assess the length of rest periods and thus the return times of some eruptions or even the conditions for magma reservoir recharge and eruption initiation (Druitt et al. 2012) (see Figure 1.9).

The work of reconstructing the history of a volcano requires a lithological and stratigraphic analysis of the emitted products, as well as the most comprehensive dating of the events, to determine the timing of the eruptions. It is essential to be able to determine the frequency of eruptions in the past and the time interval between major eruptions of explosive volcanoes, often Plinian eruptions. In Santorini, a dozen major Plinian eruptions have occurred over the last 360,000 years, with a recurrence period of about 30,000 years. The rhythmicity of these events suggests that an eruption such as the one Santorini experienced around 1600 BCE, which likely destroyed the Minoan civilization, is unlikely in the coming centuries (Druitt et al. 1989; Jenkins et al. 2015; Barberi and Carapezza 2019). The most intense volcanic event that is considered likely to occur today is a sub-Plinian eruption (Vougioukalakis et al. 2017).

Vesuvius is one of the three historically active volcanoes of Campania, one of the central provinces of Italy. Its volcanic activity shows a great variability of eruptive regime, from Plinian paroxysms such as the eruption of 79 CE to much more modest lava effusions. The last period of activity from 1631 to 1944 (see Figure 1.10) allows us to characterize these lower energy eruptions. Vesuvius has not erupted since 1944. Throughout its eruptive history, this volcano has often experienced long periods of rest lasting several centuries or tens of centuries. The longer the period of rest before the eruption, the more violent the "reawakening". The assessment of the future activity of Vesuvius should not be only based on the recent period but should integrate several cycles of activity.

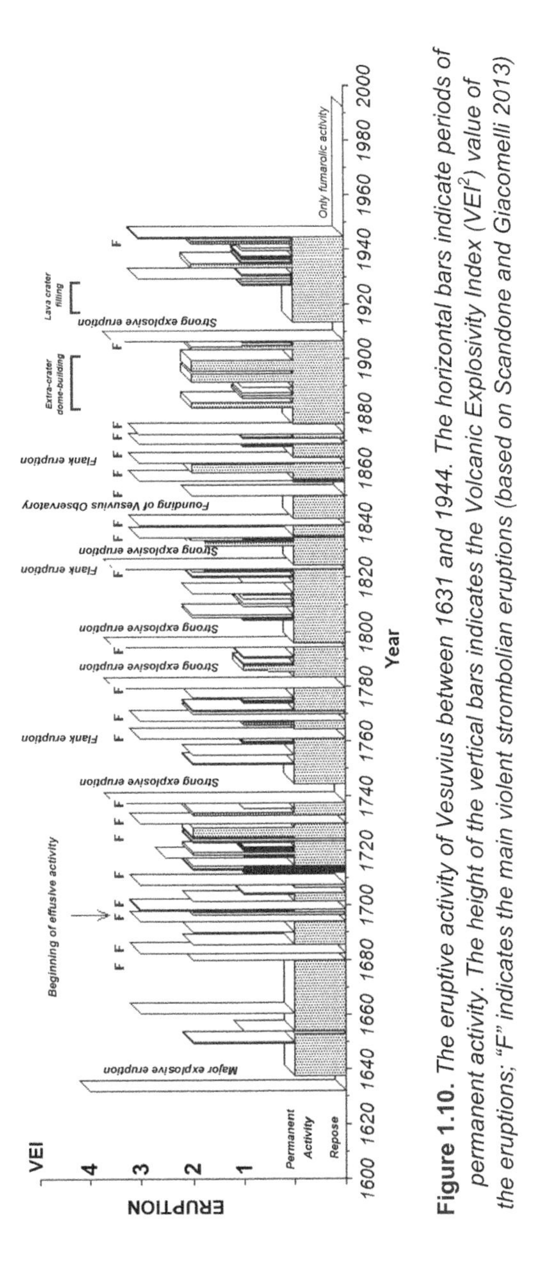

Figure 1.10. *The eruptive activity of Vesuvius between 1631 and 1944. The horizontal bars indicate periods of permanent activity. The height of the vertical bars indicates the Volcanic Explosivity Index (VEI[2]) value of the eruptions; "F" indicates the main violent strombolian eruptions (based on Scandone and Giacomelli 2013)*

2 VEI or Volcanic Explosivity Index is an index proposed by Newhall and Self (1982). A semi-quantitative scale compares the size of eruptions, with a scale of 0–8 taking into account both the magnitude of the eruption with the volume and its intensity with the height of the plume.

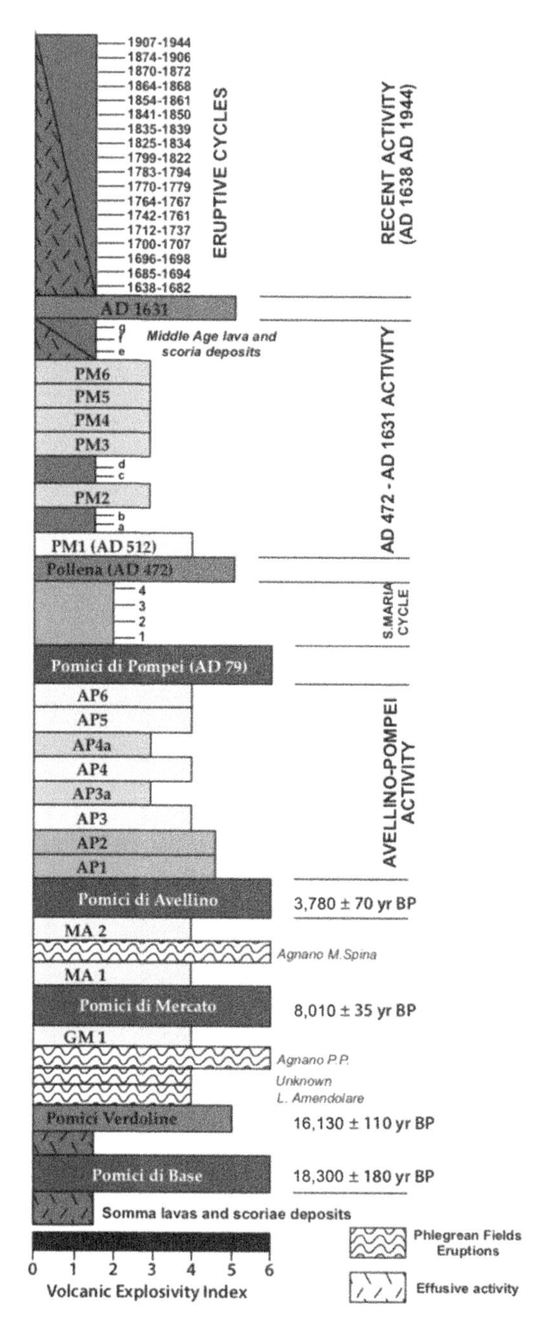

Figure 1.11. *Chronostratigraphy of the eruptions of Vesuvius in the last 18,000 years (according to Cioni et al. 2003)*

The eruption of March 1944 began with an essentially effusive activity, before a paroxysmal phase during which the eruptive column could reach nearly 10 km in altitude. It ended a period of 300 years of intense eruptive activity that had begun immediately after the sub-Plinian eruption of 1631. During this period from 1631 to 1944 (see Figure 1.10), Vesuvius was in near constant activity, with alternating effusive eruptions, and weak to violent strombolian explosions (Scandone et al. 2008; Scandone and Giacomelli 2013).

The reconstruction of the geological history of Vesuvius during the last 18,000 years (Cioni et al. 2003) (see Figure 1.11) shows a succession of large Plinian eruptions (VEI > 5), of which the eruption of 79 CE is the last. Several sub-Plinian eruptions (VEI 4–5) have marked the inter-Plinian periods, the two most recent occurring in 472 and 1631 (Arrighi et al. 2001). Alternating with these major eruptions, several smaller explosive (VEI ≈ 3) and effusive eruptions occurred. Although it is likely that not all low-magnitude eruptions have been recorded, this historical record provides an accurate picture for the activity of the Somma-Vesuvius complex over the past 20,000 years.

Figure 1.12. *Fresco in a villa in Pompeii showing Vesuvius before the eruption of 79 CE (the oldest representation of Vesuvius)*

Plinian eruptions of the type of 79 CE are rare events, spaced in time by thousands of years. The major risk at Vesuvius in the coming years is probably not related to an eruption of this type. A sub-Plinian eruption such as those of 1631 and 472 are therefore used as a reference by the Osservatorio Vesuviano and the Italian civil protection to draw up a possible eruption scenario. The statistical processing of these data indicates, for Vesuvius, that the most likely event (probability >70%) is an event of lower energy (VEI = 3). However, a sub-Plinian explosive eruption like the one in 1631, with VEI = 4 (27% probability) is considered as the reference scenario (Marzocchi et al. 2004; Gurioli et al. 2010).

1.4. Historical activity through texts, iconography and archeology

The reconstruction of the eruptive history of a volcano is based on different complementary studies. In addition to the study of deposits left by eruptions, the study of historical documents (texts, paintings, sketches, photos, etc.), when they are available, makes an important contribution to the reconstruction of the history of a volcano. It often provides a wealth of information to characterize the activity of the volcano in the near past. With the exception of the volcanoes of the Mediterranean area for which the historical period covers several millennia, the period covered is often short, a few hundred years at most. The information provided by the historical documents depends not only on this duration, on the existence of usable documents, but also on the type of activity of a given volcano. If we compare the historical periods of Piton de la Fournaise on Reunion Island and La Soufrière on Guadeloupe, two islands for which the "historical period" is more or less the same (i.e. since 1635–1640), the large number of eruptions and morphological changes described for Piton de la Fournaise provides significant information on the evolution of its eruptive and tectonic activity, whereas for La Soufrière, the historical documents only provide information on phreatic events, which represent a minor type of manifestation for this volcano.

Numerous texts, paintings, sketches and drawings describe past eruptions of Italian volcanoes, beginning with Pliny the Younger's famous description of the paroxysmal phase of the Vesuvius eruption in his two letters to the historian Tacitus. Pliny's description of a sustained eruptive event of long duration (19 hours) provides the temporal framework used by Sigurdsson et al. (1982, 1985) to place the various successive pyroclastic units in time.

Differently, the morphology of Vesuvius before the eruption of 79 CE is detailed by Strabon (58 BCE–21 CE), a Greek geographer and historian, who describes a Vesuvius whose slopes are occupied by dwellings and farmland, while the flat and desolate summit reveals its volcanic nature (see Figure 1.12). The activity of Vesuvius, like that of Etna (see Figure 1.13) or Stromboli, is widely described, regardless of the time period considered (Barberi et al. 1993; Scandone et al. 1993; Branca and Del Carlo 2004; Branca and Abate 2019).

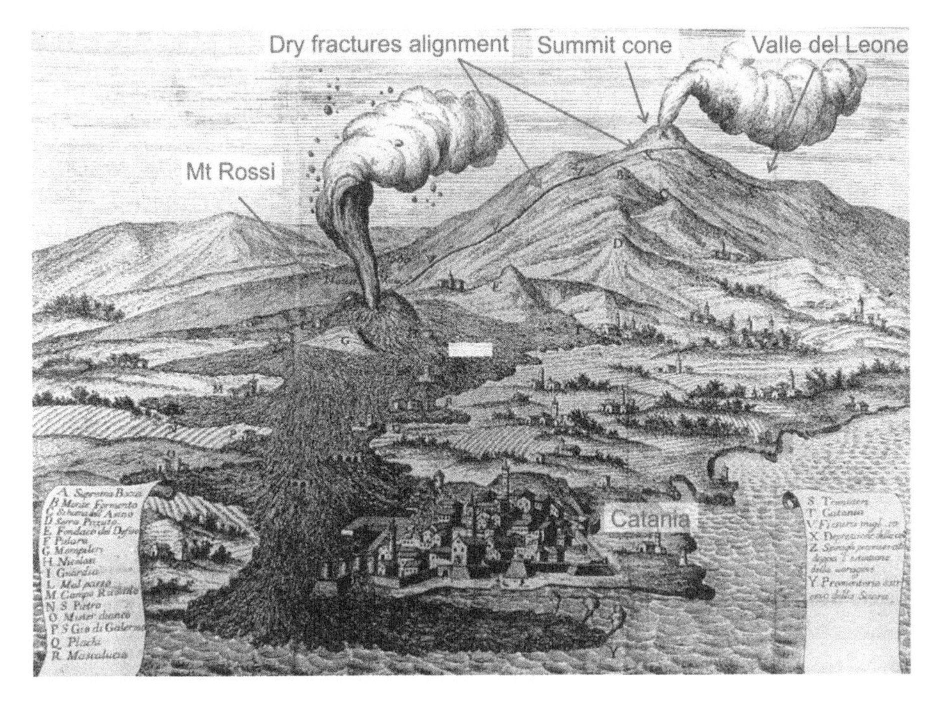

Figure 1.13. *Lava flow of Etna, coming from Mt. Rossi, entering the city of Catania in 1669. Copper engraving (1792) by Alfonso Borelli (1670) (from Branca and Abate 2019)*

Historical chronicles can sometimes replace or be extended by oral tradition (Cashman and Giordano 2008 and references therein), such as in Kilauea where Hawaiian oral tradition complements written historical chronicles dating back only to 1794 (Swanson 2008), or for countries where a rich oral tradition holds greater sway than the written word such as in the Comoros (Allibert 2015) or Indonesia (Troll et al. 2015).

1.5. The work of the pioneers

The descriptions left by geologists, volcanologists, geographers or scientists of the 18th and 19th centuries deserve an attentive reading. They relate, often with great precision, facts, events or landscapes today inaccessible to observation. We will highlight here three prominent figures who have each marked the history of volcanology: Professor Alfred Lacroix, the naturalist Jean-Baptiste Bory de Saint-Vincent and the ambassador William Hamilton. The work of another prominent figure, Thomas Jaggar, founder of the Hawaiian Volcano Observatory, will be discussed in Chapter 4.

1.5.1. *Alfred Lacroix*

Alfred Lacroix's descriptions (see Figure 1.14) of the eruptions of Mount Pelée in Martinique in 1902–1903, or those of Piton de la Fournaise on Reunion Island, are essential writings for understanding the eruptive history of these volcanoes. In May 1902, after the volcanic eruption in Martinique that killed 28,000 people, the Academy of Sciences and the Ministry of Colonies decided to send a mission to the West Indies to study the circumstances of the disaster. It arrived on June 23 and was led by Alfred Lacroix. Following a new deadly eruption on August 30, Lacroix, who had just returned to Paris, made a new trip to Martinique and stayed there until March 1903. At a time when volcanology was still in its infancy, Lacroix made essential observations for the understanding of volcanoes by describing eruptive manifestations that were unknown until then, such as the devastating pyroclastic flows, as well as the growth and destruction of the domes and spines that marked this eruption.

In 1911, he studied the Piton de la Fournaise on Reunion Island, a volcano of a different type from those of the West Indies. He drew up an inventory of the eruptions, which remains today the only source of knowledge for many eruptions of the 19th century and the beginning of the 20th century. He also described the evolution of the Dolomieu crater at Piton de la Fournaise between 1911 and 1938 (see Box 1.2), a period during which a major collapse took place, comparable in many ways to the one that, 70 years later, in 2007, was the object of much scientific attention (Michon et al. 2007, 2011; Peltier et al. 2009b; Staudacher et al. 2009). Lacroix worked hard for the creation of volcanological observatories that he had endowed with substantial resources. It is therefore partly thanks to him that France has one of the most efficient networks for volcanic risk mitigation.

A professor at the National Museum of Natural History in Paris, he wrote, among others, *Minéralogie de la France et de ses Colonies*, as well as *Minéralogie de Madagascar*, in which he described numerous mineral species, but in volcanology his major contributions are *La Montagne Pelée et ses éruptions*, published in 1904, and *Le Volcan actif de l'île de La Réunion et ses produits*, published in 1936[3].

Figure 1.14. *Alfred Lacroix (1863–1948), renowned mineralogist and geologist*

3 See http://roches.mnhn.fr/bio/lacroixbio.htm.

1.5.2. *Jean-Baptiste Bory de Saint-Vincent*

Jean-Baptiste G.M. Bory de Saint-Vincent (1778–1846) was a great traveler who visited the volcanic islands, taking advantage of the great expeditions around the world that marked the 17th century. Bory de Saint-Vincent (see Figure 1.15) was a naturalist and therefore, at that time, a polyvalent: botanist, geographer and volcanologist. He described in great detail the Piton de la Fournaise, of which he made the first ascent of the summit on October 25, 1801. His descriptions of its summit area allow us to become aware of the activity of this volcano in the 18th century. Marked by the existence of a lava lake at the summit crater and the formation of a vast lava field, probably during long-lasting eruptions (Lénat et al. 2001), this type of activity differs significantly from the one we know today for this volcano (see Box 1.2).

He opened the way to scientific research on the Piton de la Fournaise. His work *Voyage dans les quatre principales îles des mers d'Afrique*, published in 1804, contains numerous observations of the central zone of that volcano, and also of the littoral zone with its numerous recent flows, today buried under vegetation or by other more recent flows.

Figure 1.15. *Bory de Saint-Vincent (1778–1846)*

The collapse, over 300 m deep, of the summit caldera of Dolomieu at Piton de la Fournaise in April 2007, has revealed formations that were previously inaccessible to observation. This natural section, revealing the recent history of Piton de la Fournaise, has aroused great interest and led to a re-evaluation of the history of the volcano's summit over the past three centuries.

The approach combines a lithostratigraphic interpretation of these new outcrops and the crater periphery, made possible by the acquisition of high-resolution photographs, with a detailed examination of historical (post-1640) documents in which the morphology or eruptive activity of the summit zone is described (Peltier et al. 2012; Michon et al. 2013).

This comparison of iconographic documents and geological observations (Figures 1.17–1.19) has allowed us to make a new contribution to the knowledge of the history of the building of the terminal cone of Piton de la Fournaise. The changes in the eruptive dynamics of the volcano are described, with the construction of an ancient cone during which phases of explosive activity dominated, followed by a more effusive period, itself subdivided into two periods. Links with the successive collapses of the summit crater (or caldera) can also be established. The current phase of activity following the collapse of 2007 can thus be evaluated by considering this past functioning.

These studies show, in particular, that the activity of Piton de la Fournaise has recently evolved. During the 18th and 19th centuries, the eruptive activity was characterized by several sustained, long-lasting, effusive eruptions, associated with frequent phreatic explosions and ending with phreatomagmatic explosions. On the other hand, the 20th century and now the 21st century correspond to periods mainly marked by periodic effusive activity, coexisting with minor phreatic explosions during summit collapses resulting from lateral eruptions.

The current central cone of Piton de la Fournaise was probably built quite rapidly, during a phase of sustained effusive activity (Lénat et al. 2001), essentially centered on the western summit crater (Bory) and following a period of intense explosive activity. Since the end of the 19th century, the growth of the central cone is less, and the activity has moved eastward (Dolomieu), where frequent caldera or pit-crater collapses and phases of refilling of these craters follow each other.

Box 1.2. *The history of the morphological evolution of the Dolomieu crater – Piton de la Fournaise, Reunion Island*

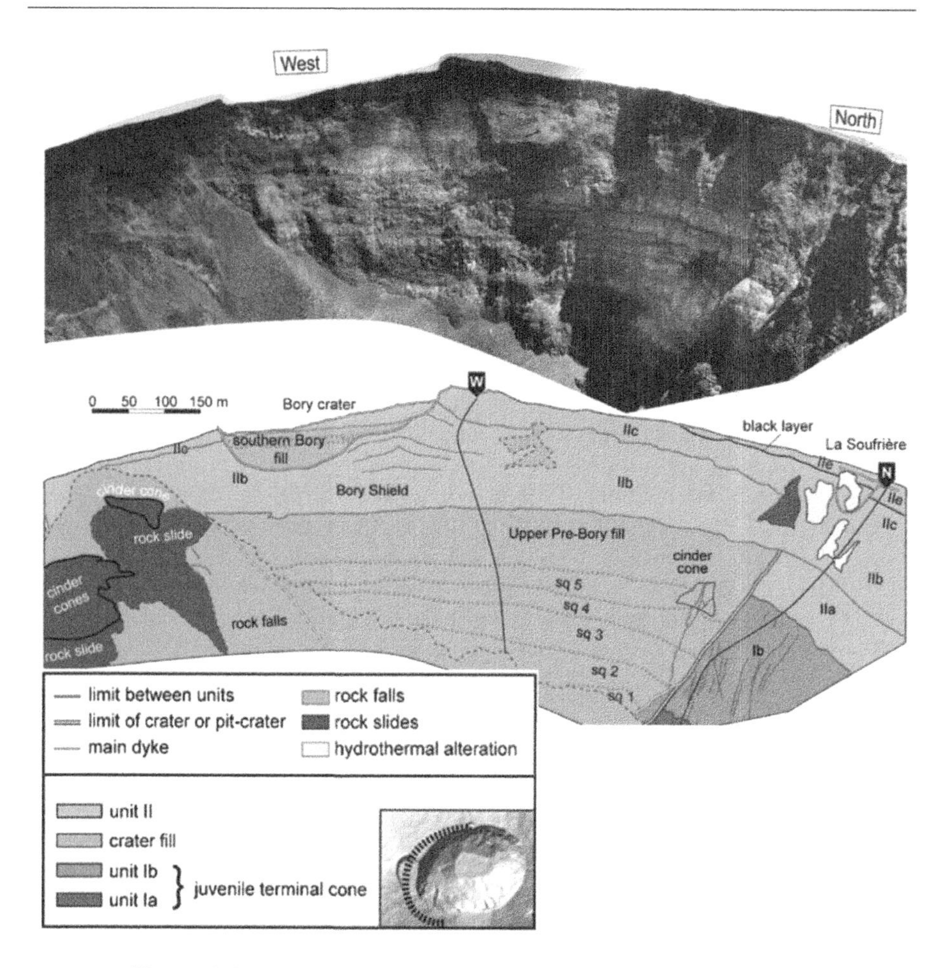

Figure 1.16. *Panorama and interpretation of the formations in the western wall of the Dolomieu crater. For a color version of this figure, see www.iste.co.uk/lenat/hazards.zip*

COMMENT ON FIGURE 1.16.– *Mosaic of photographs taken in September 2009. Units in green correspond to emissions from the Bory and Dolomieu craters. The blue units represent the refilling of an older crater, revealed by the 2007 collapse (from Peltier et al. 2012).*

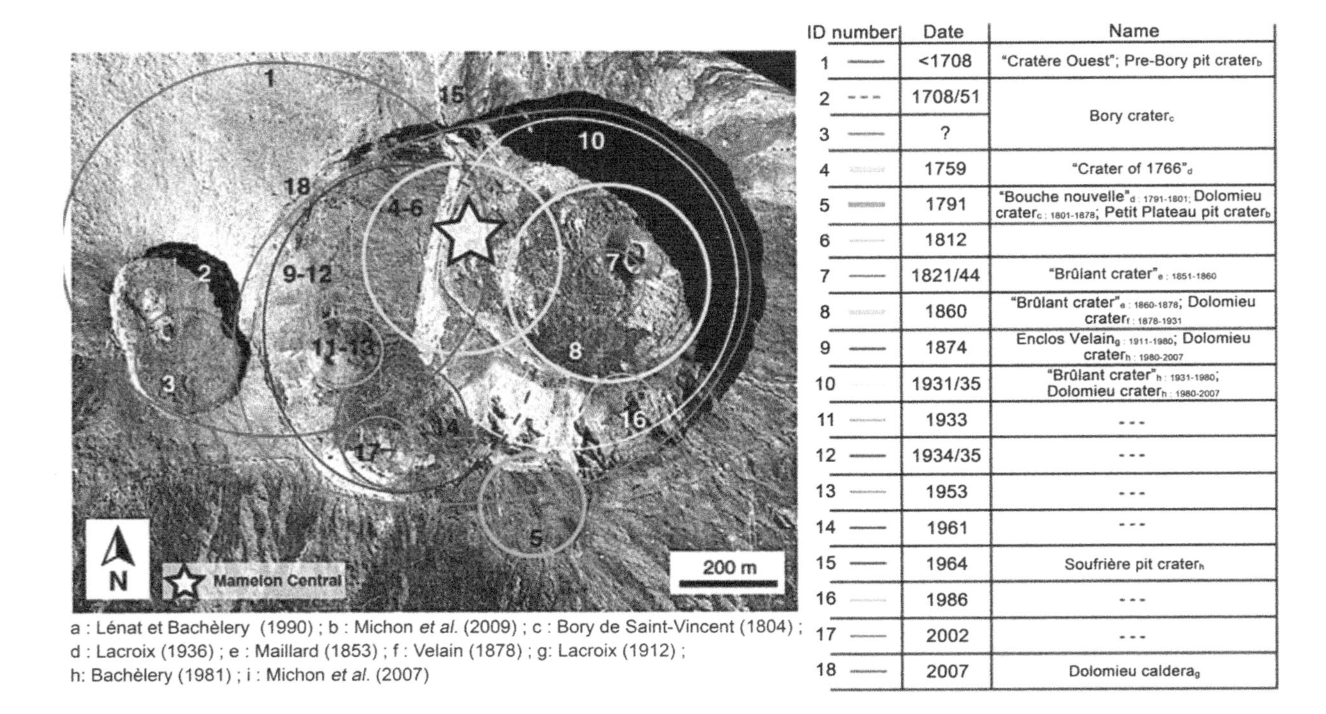

ID number	Date	Name
1	<1708	"Cratère Ouest"; Pre-Bory pit crater$_b$
2	1708/51	Bory crater$_c$
3	?	
4	1759	"Crater of 1766"$_d$
5	1791	"Bouche nouvelle"$_{d\ :\ 1791\text{-}1801}$; Dolomieu crater$_{c\ :\ 1801\text{-}1878}$; Petit Plateau pit crater$_b$
6	1812	
7	1821/44	"Brûlant crater"$_{e\ :\ 1851\text{-}1860}$
8	1860	"Brûlant crater"$_{e\ :\ 1860\text{-}1878}$; Dolomieu crater$_{f\ :\ 1878\text{-}1931}$
9	1874	Enclos Velain$_{g\ :\ 1911\text{-}1980}$; Dolomieu crater$_{h\ :\ 1980\text{-}2007}$
10	1931/35	"Brûlant crater"$_{h\ :\ 1931\text{-}1980}$; Dolomieu crater$_{h\ :\ 1980\text{-}2007}$
11	1933	---
12	1934/35	---
13	1953	---
14	1961	---
15	1964	Soufrière pit crater$_h$
16	1986	---
17	2002	---
18	2007	Dolomieu caldera$_g$

a : Lénat et Bachèlery (1990) ; b : Michon *et al.* (2009) ; c : Bory de Saint-Vincent (1804) ; d : Lacroix (1936) ; e : Maillard (1853) ; f : Velain (1878) ; g: Lacroix (1912) ; h: Bachèlery (1981) ; i : Michon *et al.* (2007)

Figure 1.17. *Delineation of the successive collapses at the summit of Piton de la Fournaise, identified from historical documents and scientific literature. The overlapping of the different collapses is clearly visible. The area collapsed in April 2007 (orange line) includes all the collapses of the last 300 years (from Michon et al. 2013). For a color version of this figure, see www.iste.co.uk/lenat/hazards.zip*

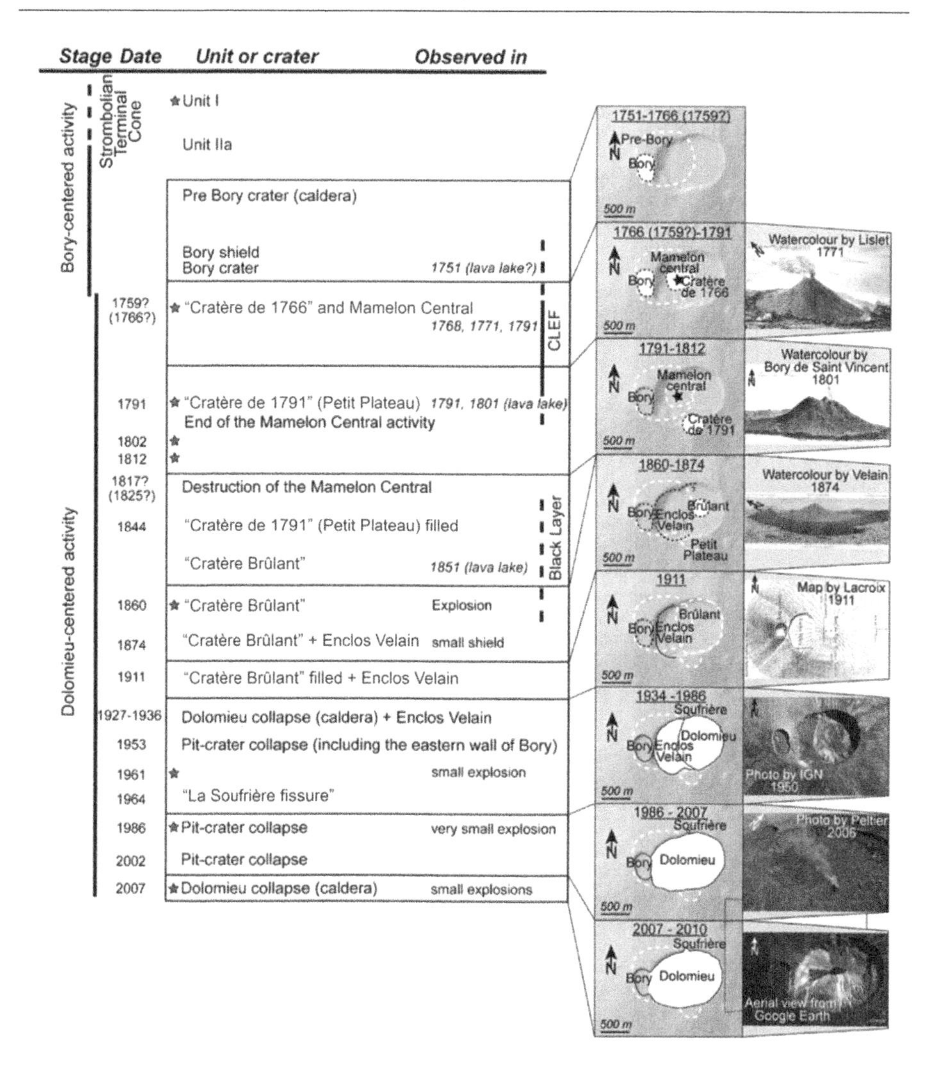

Figure 1.18. *Summary of the historical activity of the Piton de la Fournaise, compared with the morphological evolution of the summit of the Piton de la Fournaise deduced from available iconographic documents. For a color version of this figure, see www.iste.co.uk/lenat/hazards.zip*

COMMENT ON FIGURE 1.18.– *The red stars represent the most explosive phases. Distinct periods define the major changes. White dotted lines and white areas define filled or active craters for each period (from Peltier et al. 2012). Reproduced watercolors and drawings are from Bory de Saint-Vincent (1804), Vélain (1878) and Lacroix (1936, 1938).*

1.5.3. *William Hamilton*

Nothing predestined William Hamilton (1731–1803) (see Figure 1.19), a Scottish aristocrat and British ambassador to the Kingdom of Naples, to become one of the world-renowned volcanologists of his time. From 1765 to 1779, he witnessed several eruptions of Vesuvius and faithfully described their progress. He reported on these observations through regular publications in the *Philosophical Transaction of the Royal Society of London*, the first of which in 1768 (Hamilton 1768) marked his entry into the scientific world.

Figure 1.19. *Sir William Hamilton by David Allan*

In 1776, he published his contributions in a book entitled *Campi Phlegraei*, accompanied by numerous illustrations in gouache by Pietro Fabris. This work of great scientific interest allows us to visualize the morphological evolution of Vesuvius and its crater and to follow the sequence of eruptions of this volcano, which today we describe as violent strombolian to effusive. This work was completed with a supplement in 1779 relating the exceptional eruption of that same year (see Figure 1.20).

Figure 1.20. *The eruption of Vesuvius on the morning of August 9, 1779. An impressive eruptive column rises above the crater. Reproduction of a gouache by Pierre Fabris in Pausilippe*

1.6. The contribution of old maps

The use of old maps also provides a lot of information on the transformations that volcanoes have undergone over time. Their accuracy increases, of course, as the means of cartography improve. The imprecision of the oldest documents sometimes requires a lot of precaution before they can be used. Old maps provide information about how people of the time perceived their space and the importance they gave to places. In this respect, volcanoes often take an important place because they are a strong element of the landscape and of the life of the inhabitants. The maps constitute milestones in the chronology of events structuring a landscape, marking the main changes in geography, and also depending on the evolution of techniques of perception and representation of space (see Figure 1.21). On Reunion Island, this evolution is particularly noticeable. Iconographic and cartographic representations of Piton de la Fournaise evolve over time, moving from a cartography of discovery of then-unknown spaces, struggling to precisely circumscribe the location and morphology of the volcano, to a cartography constructing a conceptual image of the volcano, refined as mapping techniques advance (Germanaz 2005, 2016).

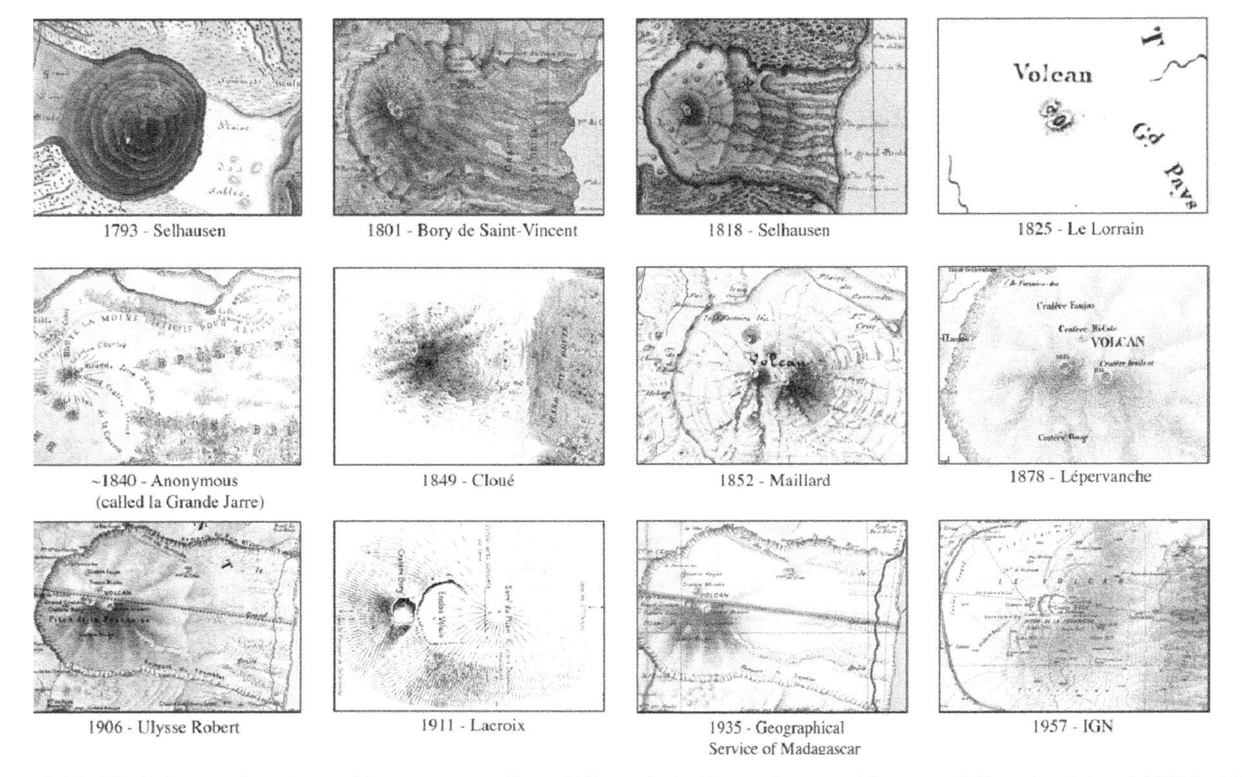

Figure 1.21. *Evolution and cartographic representation of Piton de la Fournaise on old maps of Reunion Island (1793–1957). This period clearly shows the growth of the Dolomieu crater (formerly "Cratère Brûlant") at the expense of the Bory crater (document Germanaz 2016). For a color version of this figure, see www.iste.co.uk/lenat/hazards.zip*

From being purely descriptive of the landform, the 18th century planimetric maps quickly evolved toward the first geological maps representing the most striking flows of the time, whether at Etna (Branca and Abate 2019) or on Reunion Island (see Figure 1.22) and the Comoros (Bory de Saint-Vincent 1804; Lacroix 1936, 1938). They are an important tool in volcanic risk management (Leone and Lesales 2006).

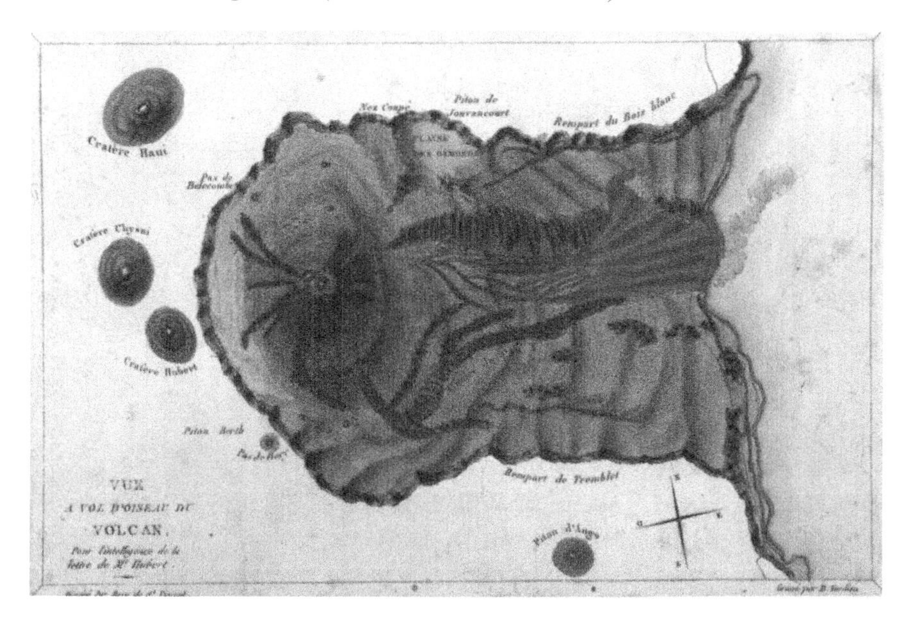

Figure 1.22. *Cartographic representation of the eruption of the year X (1801–1802) at Piton de la Fournaise (Bory de Saint-Vincent 1804). For a color version of this figure, see www.iste.co.uk/lenat/hazards.zip*

1.7. Volcanic archeology

The behavior of populations in the face of natural disasters (the combination of their knowledge of hazards, their perception of risk, the past experiences and the way situations make sense, and their compliance with instructions) is a determining aspect in terms of risk. Volcanoes are no exception to this logic, and there is a growing body of work in the humanities and social sciences on these issues (Morin et al. 2016; Fearnley et al. 2018; Avvisati et al. 2019). Questions regarding the behavior of populations faced with eruptions can also be addressed from past eruptions, in what can be called volcanological archeology. The examples are numerous. We refer to the work of Cashman and Giordano (2008) for a compilation as well as the special issue of the journal *Quaternary International* (Sevink et al. 2019 and references cited).

In recent years, the Huaynaputina eruption of February-March 1600 in southern Peru (VEI 6) has been the focus of a major archeological project. This eruption was one of the largest historical Plinian eruptions and the most voluminous eruptive event in the history of South America, causing a global cooling of 1.1° and the burial of numerous Inca villages (more than 12 already identified), in an area 75 km east of Arequipa. The Plinian phase produced about 15 km^3 of pumice fallout over an area covering more than 400,000 km^2, according to field data and the interpretation of Spanish chronicles (Thouret et al. 2002; Prival et al. 2020). Archeologists have been able to uncover evidence that the area was inhabited at the time of the eruption, based on fragments of ceramic pots and vessels found within the fallout. Here, volcanologists are working with archeologists to uncover these lost villages. Various geophysical techniques such as geo-radar, electromagnetic methods and thermal imagery allowed to identify the Inca infrastructures buried under the pyroclasts.

The research undertaken in Pompeii, in the ruins of the city destroyed by the eruption of Vesuvius in 79 CE, now has a completely different scope since it allows us to link the different phases of the eruption, deduced from the study of the deposits and historical documents, to the situation and possible attitude of the victims buried under the pumice fallout.

Were the inhabitants of Pompeii in 79 CE unlucky? Yes, certainly. We have seen (see section 1.4.2) the low frequency of Plinian eruptions, such as that of 79 CE. Moreover, the geological archives left by past eruptions of Vesuvius have shown that the fallout from explosive eruptions, since the Plinian eruption of Avellino about 3,700 years ago, has spread mainly east and northeast (Andronico and Cioni 2002; Sulpizio et al. 2010), with the prevailing winds in the region being mainly westerly. The deposits left by the first phase of the 79 CE eruption show an unusual extent to the southeast and south, thus directly concerning the city of Pompeii. The 79 CE eruption had two main phases: first, a Plinian eruptive column that caused widespread tephra fallout; then, a column collapse phase that generated pyroclastic density currents (Sigurdsson et al. 1985). Although both phases caused casualties, the eruption was not continuous, which could have caused the inhabitants still present in Pompeii during the first phase to leave their homes (Scarpati et al. 2020). Nearly half of the victims were in the streets and on the roads during the second phase of the eruption (Luongo et al.

2003a, 2003b). Important excavation work has been undertaken in ancient Pompeii, allowing, among other things, to propose an alternative for the exact day of the eruption of 79 CE (October 24 instead of August 24). This work also showed that a significant proportion (≃38%) of the victims died in their homes, during the first phase of the eruption, due to the collapse of the flat roofs of their houses under the weight of the pyroclastic deposits accumulated in large quantities due to that peculiar wind regime (Luongo et al. 2003b). This observation led the authorities in charge of civil protection to modify the so-called "red zone" delimited in case of a major eruption.

1.8. Eruptive dynamics, types of eruptions, structural evolution: the use of volcanic "archives" through geological field interpretation

The reconstruction of the history of volcanoes over long periods of time implies, of course, a geological analysis of the products emitted by past eruptions. This usually begins with the interpretation of the geological sequences and the characterization of the processes that led to their formation. Magmatic or igneous rocks are present in a wide variety of lithological facies. They are not only the result of a simple cooling and crystallization process but also the product of flow, fragmentation, sedimentation and reworking phenomena that often reflect complex eruptive regimes. Field geology requires its own learning, especially for volcanic rocks that often constitute geological units of small extent, with multiple facies changes, sometimes within the same unit. Contrary to what is often assumed, it is often more difficult to correctly interpret a sequence of pyroclastic fallout and flow that can be partially reworked (see Figure 1.23) than to mathematically solve a physical problem. This implies observation and interpretation skills that must be based on a learning and knowledge that, unfortunately, is often far too absent in our university teaching.

The field work must allow the characterization of the geological formation in situ, in its context and with respect to its environment. This is a fundamental prerequisite for any laboratory analysis carried out on the samples collected. This fieldwork will also allow us to establish the relationships between the units and their relative chronology. This is obviously an essential aspect of the reconstruction of the eruptive history of a volcano. "If the knowledge of the field geology is poor, all studies based on collected samples and field measurements will also be poor. Conversely, a good understanding of the field geology provides the basis for good geologic interpretation" (Jerram and Petford 2012).

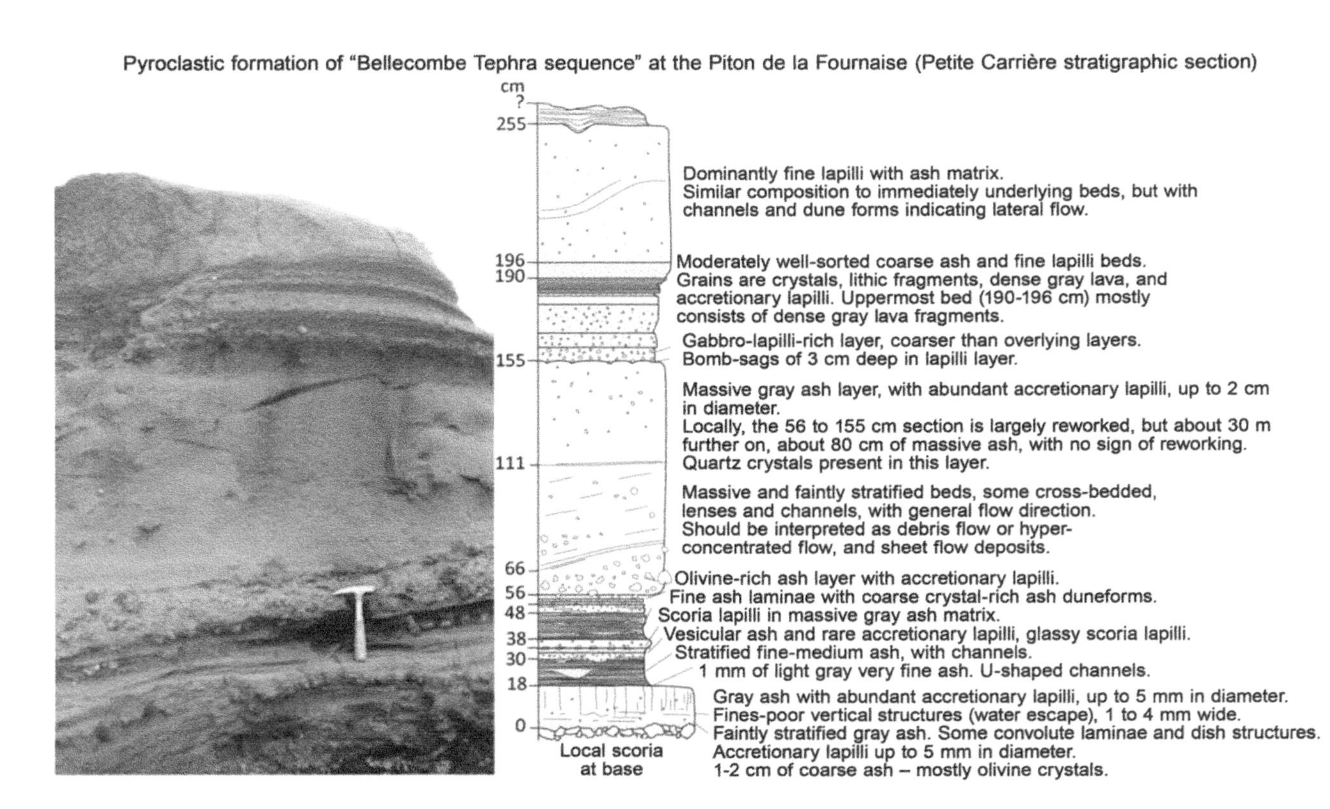

Figure 1.23. *a) Photo of deposits and b) stratigraphic column of one of the main outcrops of the "Bellecombe Tephra sequence" at the Piton de la Fournaise in on Reunion Island*

COMMENT ON FIGURE 1.23.– *This deposit is mainly the result of phreatomagmatic explosions that produced breccias, pyroclastic flows, and ash and lapilli fallout. The upper part of the Bellecombe Tephra sequence contains numerous hydrothermally altered fragments indicating the involvement of a mature, deep-seated hydrothermal system (photo: P. Bachèlery, after Ort et al. 2016).*

Volcanic rocks, derived from magmas emitted on Earth's surface, are directly the result of eruptions. These rocks are therefore the essential keys to decipher the eruptive history, even if intrusive or plutonic rocks, emplaced at various depths, are also part of this history. The study of the deposits on land is often extended by the study of these same deposits at sea, in order to obtain the most accurate knowledge of the whole geological record.

The work carried out on La Soufrière on Guadeloupe shows how much knowledge of the activity of this volcano has progressed since the phreatic eruption of 1976. Studies of deposits on land (Boudon et al. 1988) and at sea (Deplus et al. 2001; Boudon et al. 2007) have characterized the volcanic and tectonic activity of the volcano over the last 9,000 years, highlighting the exceptional recurrence of catastrophic flank landslides, all related to magmatic activity (Komorowski et al. 2005; Boudon et al. 2007; Komorowski et al. 2008). Today, a detailed chronology of the eruptive history of the Grande Découverte-La Soufrière complex on Guadeloupe exists even for the last 50,000 years (Legendre 2012; ANR CASAVA Final Report) (see Figure 1.24).

The chronostratigraphic data, in particular the characterization of turbulent and dilute pyroclastic flow deposits, allow to characterize not only the frequency and the magnitude of the magmatic activity, but also the chain of events (dome growth, explosive phases, flank destabilization and directed explosions). This shows that most eruptions consist of multiple phases with time-varying eruptive regimes, following an eruptive scenario that could be analogous to that of the 1980 Mount St. Helens eruption (Lipman and Mullineaux 1981) or the Soufrière Hill eruption on Montserrat (Sparks and Young 2002; Voight et al. 2002).

Examples of geological reconstructions of volcano activity are numerous in the literature. In a different context from that of the West Indies, one can cite the evidence of alternating Strombolian and violent Strombolian dynamisms at Antuco volcano in Chile (Romero et al. 2020) or the evidence of pyroclastic flows at Stromboli (Lucchi et al. 2019). This work of "reading" the volcanic archives allows us to recognize the different "modes of expression" of the volcano, the types of eruptions it has experienced in the past, the structural and dynamic characteristics and the temporal relationships between these different types of eruptions.

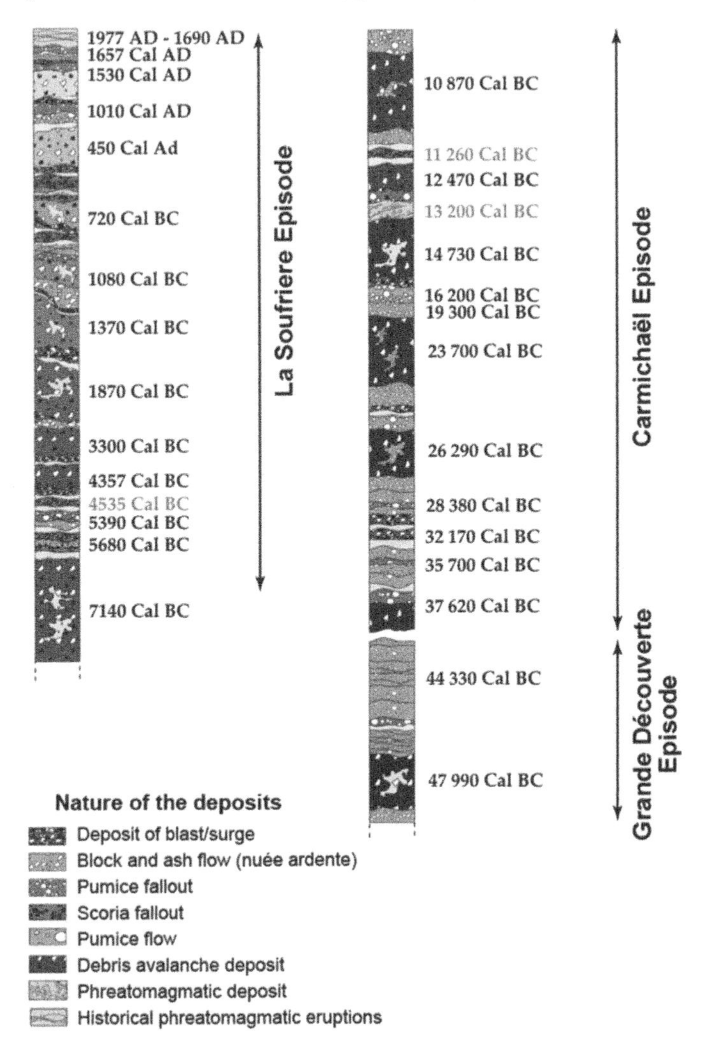

Figure 1.24. *Synthetic log of the eruptive stratigraphy of the Grande Découverte-La Soufrière complex over the past 50,000 years. Shown in green are the three explosive events of the Madeleine-Trois-Rivières complex (Legendre 2012). Extract from the final report of the ANR CASAVA*[4] *project. For a color version of this figure, see www.iste.co.uk/lenat/hazards.zip*

4 https://sites.google.com/site/casavaanr/.

On a basaltic volcano like Piton de la Fournaise, the operational monitoring strategy is going to be adapted according to the different types of eruptions the volcano has experienced in the past. Although explosive paroxysms are known (Bachèlery 1981; Morandi et al. 2016) (see also Figure 1.23), the reconstruction of the eruptive history of Piton de la Fournaise shows that the most common eruption type for this volcano in the current period is the emission of lava flows from lateral fissures, opening on the flanks, and emitting small amounts of pyroclastic material (Bachèlery 1981; Lénat and Bachèlery 1988; Villeneuve and Bachèlery 2006; Peltier et al. 2009a; Staudacher et al. 2016) (see Figure 1.25).

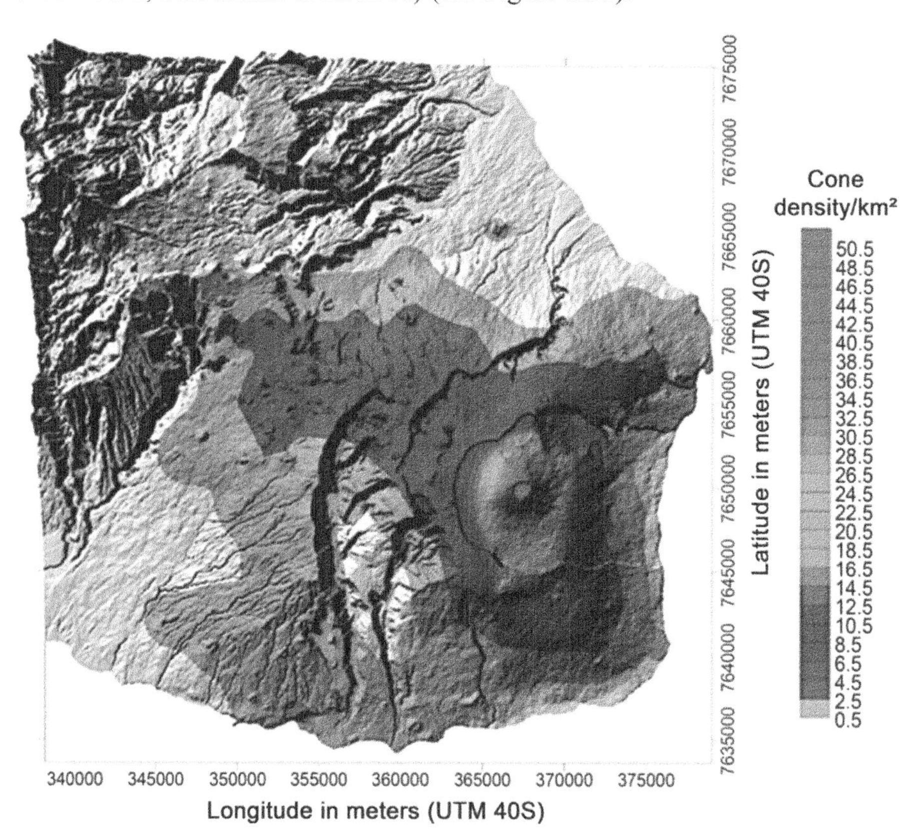

Figure 1.25. *Spatial distribution of historic and prehistoric eruptive cones and fissures on the Piton de la Fournaise volcano (extract from Villeneuve and Bachèlery 2006). For a color version of this figure, see www.iste.co.uk/lenat/hazards.zip*

1.9. Structural framework and evolution

The structural evolution of a volcanic edifice during the Holocene is, of course, part of its recent geological history. Two types of tectonic events are particularly important in the monitoring of currently active volcanoes, because their frequency is such that they may occur with a non-zero probability in the near future: flank landslides, which concern many volcanoes, and summit collapses forming a caldera or a pit-crater.

The importance of flank destabilizations has been highlighted at many eruptive sites regardless of their nature, based on consideration of the geology of the formations on land (Martí 2019) and also from work at sea (Le Friant et al. 2015). These events are now fully integrated into monitoring strategies.

In the case of destabilization resulting in a sector collapse, such as that experienced by Mount St. Helens in 1980, the rocks involved in the collapse form a characteristic deposit called "debris avalanche deposit". This deposit can be recognized by a bulge in the terrestrial or submarine morphology, and by its brecciated aspect, where brecciated blocks (block facies) coexist within a brecciated matrix (matrix facies) (Ui et al. 2000; Carey and Schneider 2011; Perinotto et al. 2015). In the case of destabilization resulting from the intrusion of a viscous magma, they may be followed by a directed lateral explosion (blast). In general, these flank collapses deeply mark the volcanic morphology, leaving a characteristic horseshoe-shaped amphitheater and a large deposit that can be channeled into the valleys or spread widely in a fan at the foot of the edifice. These gravity flows, when affecting island volcanoes, are likely to generate large tsunamis (Paris et al. 2017; Pistolesi et al. 2020).

Flank collapses with debris avalanche deposits, better understood since the 1980 eruption of Mount St. Helens, are considered in potential eruptive scenarios. They have been recognized for many volcanoes in both continental and oceanic domains (Gorshkov 1959; Moore et al. 1989; Holcomb and Searle 1991; Normark et al. 1993; Carracedo et al. 1999; Day et al. 1999; Van Wyk De Vries et al. 2001; Masson et al. 2002, 2008; Mitchell 2003; Oehler et al. 2008; van Wyk de Vries and Davies 2015; Paris et al. 2018), sometimes reoccurring, whether for stratovolcanoes with differentiated magmas or for basaltic shield volcanoes. Work carried out on La Soufrière on Guadeloupe has revealed nine collapses of varying magnitude in less than 10,000 years, the last of which was in 1530

(Komorowski et al. 2008). At Mount Pelée, at least two such events are also known (Vincent et al. 1989; Boudon et al. 2013). On Reunion Island, giant flank landslides also mark the geological history of the island (Lénat et al. 1989; Oehler et al. 2008; Le Friant et al. 2011).

Figure 1.26. *The summit crater Dolomieu at Piton de la Fournaise (Reunion Island) on April 5, before the caldera collapse of April 6, 2007. The collapse, more than 300 m deep, took place mainly in less than 24 hours (photo: Lucette Ferlico)*

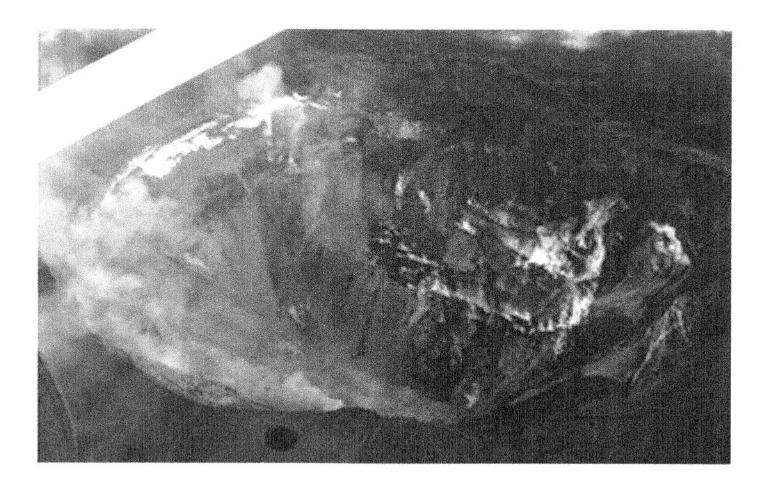

Figure 1.27. *The summit crater Dolomieu at Piton de la Fournaise on April 7. The collapse, more than 300 m deep, took place mainly in less than 24 hours (photo: Lucette Ferlico)*

While collapses affecting the summit of basaltic volcanoes and forming a so-called pit-crater are relatively common, five caldera collapses were also observed and monitored by monitoring networks during the late 20th and early 21st centuries: Fernandina, Galapagos, 1968 (Simkin and Howard 1970); Myakejima, Japan, 2000 (Geshi et al. 2002); Piton de la Fournaise, Reunion Island, 2007 (Michon et al. 2007, 2009; Peltier et al. 2009b; Staudacher et al. 2009); Bárdarbunga, Iceland, 2014 (Sigmundsson et al. 2014; Gudmundsson et al. 2016; Riel et al. 2016; Sigmundsson 2019); and Kilauea, Hawaii, 2018 (Anderson et al. 2019; Neal et al. 2019). They serve as a reference to describe and anticipate this type of major change in the morphology and structure of a volcano. Calderas are large depressions, more or less circular or elliptical, formed during the collapse of the top of the volcano following large eruptions or lateral intrusions that affect a main magma chamber located a few kilometers deep (MacDonald 1965). The collapse of the Dolomieu caldera at Piton de la Fournaise in 2007 (see Figures 1.26 and 1.27) occurred along a circular fault system and is clearly attributed to the emptying and incremental collapse of the roof of a magma chamber located at shallow depth (Michon et al. 2007, 2011; Massin et al. 2011). These events, although infrequent, must be considered in monitoring strategies. At Piton de la Fournaise, at least two collapses of the Dolomieu caldera have occurred in the last 100 years (see Box 1.2).

1.10. The use of distant archives

1.10.1. *The record of large eruptions in marine and lake sediments*

Tephrochronology, often used in the near field to determine the history of volcanoes, can also be addressed for medium- to long-distance areas. Marine (Bazin et al. 2019) and lake sediments (Lane et al. 2011, 2013), or speleothems (Bazin et al. 2019), are excellent archives for tephra and can provide a detailed record of Quaternary explosive volcanism.

Volcanic sequences described on land may be incomplete due to erosion, or made inaccessible because they are covered by younger products, or simply difficult to interpret due to dense vegetation cover, especially in the intertropical zone. Distant deposits left by explosive eruptions (tephra) in cores taken from marine or lacustrine sediments (see Figure 1.28) are then of great use to reconstruct more completely the records of explosive eruptions of nearby volcanoes, allowing access to the petrology of these events or to date them (Paterne et al. 1988, 1990; Narcisi 1996; Fretzdorff et al. 2000; Hamann et al. 2010; Sulpizio et al. 2010; Gudmundsdóttir et al. 2011; Cassidy et al. 2014; Çağatay et al. 2015; Albert et al. 2017; Leicher et al. 2019). For example, a tephra layer related to the eruption of Pavin Lake (4,720 ± 170 BCE), from a few millimeters to a few centimeters thick, is found in many peatlands and lacustrine deposits in the Massif Central (Juvigné and Miallier 2016).

Figure 1.28. *Sampling and exploitation of deep-sea sediment cores on board an oceanographic vessel*

COMMENT ON FIGURE 1.28.– *The sediment cores contain pelagic or hemi-pelagic sediments rich in foraminifera, allowing a chronology to be obtained thanks to* ^{14}C *and* $\delta^{18}O$*, gravity deposits from turbidity currents, and tephra layers from the sedimentation of eruptive plumes (photos: P. Bachèlery).*

These tephra layers are made up of the accumulation of glassy or small-size pumice fragments. The fallout from the volcanic plumes, transported by the wind and then by the marine currents for the deposits at sea, can thus constitute a thin deposit on vast surfaces. Their identification is not always easy. Problems such as bioturbation, dispersal by currents and marine erosion can disrupt the preservation of marine tephra or thin volcaniclastic layers (Gudmundsdóttir et al. 2011; Cassidy et al. 2014). Collected near volcanoes or volcanic islands, more distant tephra deposits thus provide a "history" of nearby explosive eruptions, complementing the knowledge of eruptive history determined from the study of deposits conducted in the proximal domain. Many works have demonstrated the importance of these approaches and their contribution to volcanic hazard forecasting (Watkins et al. 1978; Gehrels et al. 2006; Bertrand et al. 2008; Sulpizio et al. 2008; Insinga et al. 2014).

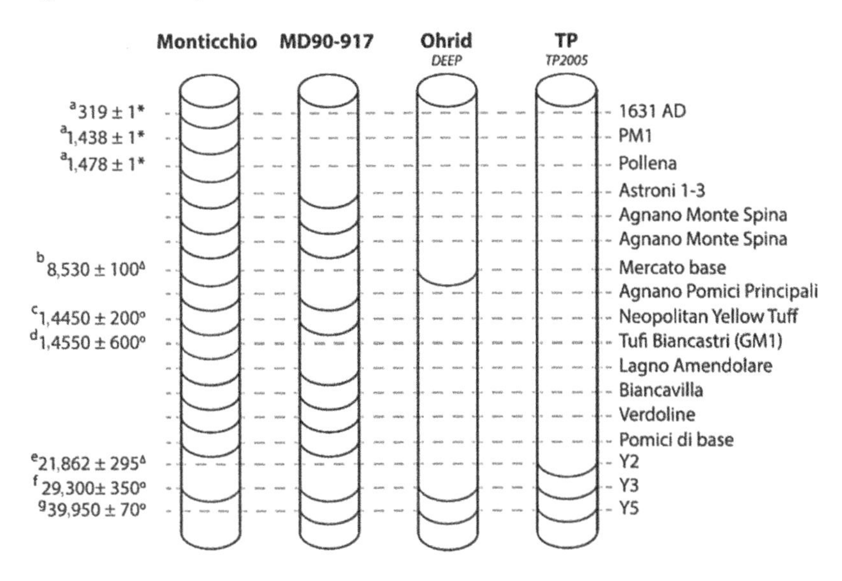

Figure 1.29. *Well-dated tephra levels identified in some terrestrial and marine sediment cores from the Mediterranean area (lake sediments: Monticchio Maar, Italy, Ohrid, Greece, and Tenaghi Philippon (TP), Turkey; marine sediments: MD90-917, southern Adriatic Sea). For a color version of this figure, see www.iste.co.uk/lenat/hazards.zip*

COMMENT ON FIGURE 1.29.– *Tephra in red are dated using $^{40}Ar/^{39}Ar$ or ^{14}C methods (from Bazin et al. (2019) – see references included).*

Another use of marine and lacustrine tephra is the correlation of (distal) deposits from very large (Plinian) eruptions and relatively infrequent explosive eruptions (Pyle et al. 2006; Bazin et al. 2019). These deposits, present over large areas, provide regional isochronous markers that can be cross-correlated over large areas. They are widely used in paleoclimate studies (Lowe 2011). While sedimentary archives, whether marine or lacustrine, can be used in many contexts, it is probably in the Mediterranean area that the greatest number of studies is available. A chronology based on major events is now established with good accuracy, from independent sources mixing data from lake sediments (varves) and data from marine sediments (Insinga et al. 2014; Bazin et al. 2019) (Figure 1.29). Tephra layers with a wide distribution are used as a reference for further work. If the Y-5 tephra is the most widespread layer in the deep-water sediments of the eastern Mediterranean (Keller et al. 1978; Pyle et al. 2006), there are several other reference layers for which the relationship with an eruption on land is more or less well-established (Zanchetta et al. 2011; Albert et al. 2015, 2017). In the Ionian, Tyrrhenian and Adriatic seas, these tephra layers serve as a basis for locating in time the main tectonic and gravity events, some of which are of direct interest to current or recent volcano activity.

In the marine realm, because volcanic environments are often tectonically active, volcaniclastic tephra deposits (of pyroclastic origin) are often associated with other types of volcaniclastic deposits, particularly on the continental slopes and flanks of volcanic islands and on the surrounding abyssal plain. These are epiclastic sediments, resulting from the degradation of lava flows and other geological units constituting the submarine or subaerial flanks of volcanoes (Manville et al. 2009; Carey and Schneider 2011; Cassidy et al. 2014). They result from the fragmentation of pre-existing rocks and can have a wide range of thicknesses and textures, depending on their origin (Carey and Schneider 2011). Volcaniclastic deposits thus originate from the transport of volcanic clasts as a result of landslides, collapses and mass flows, floods or from the entry of pyroclastic flows into the sea (Le Friant et al. 2009). They can form density currents whose deposits (turbidites) are characteristic (Bouma 1962; Piper and Normark 2009). Their diversity is inherent to the diversity of fragmentation, transport and deposition processes from which they result. For volcanic islands, they provide information about the gravity and eruptive processes that have affected the island over several hundred thousand years (Garcia and Hull 1994; Trofimovs et al. 2008; Babonneau et al. 2016; Hunt and Jarvis 2017). In sediment cores, these deposits can be interbedded with tephra fallout offering the opportunity to reconstruct the volcanic and tectonic history of a volcano or volcanic area (Schneider et al. 2001; Gudmundsdóttir et al. 2011; Köng et al. 2016; Hunt and Jarvis 2017) (see Figures 1.30 and 1.31).

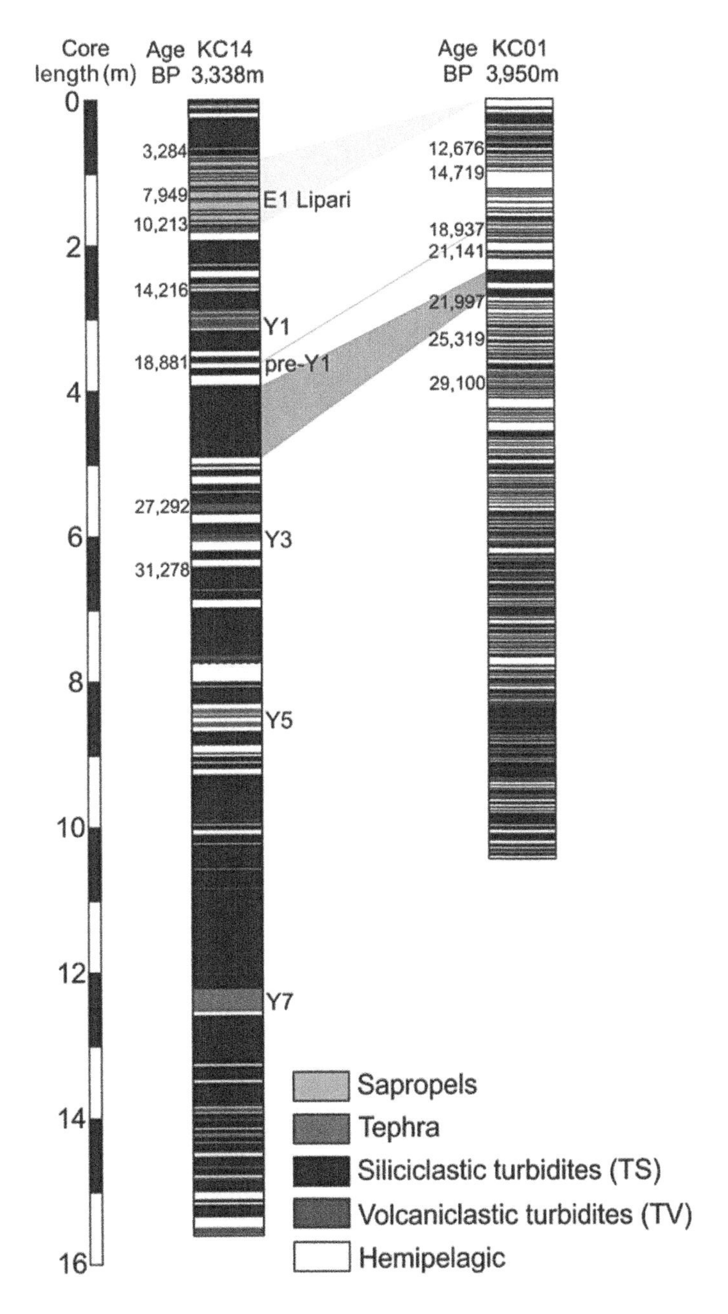

Figure 1.30. *Logs of two sediment cores taken from the Ionian Sea, down the slopes of Sicily and Calabria. For a color version of this figure, see www.iste.co.uk/lenat/hazards.zip*

COMMENT ON FIGURE 1.30.– *Major gravity events in the area, particularly volcaniclastic turbidites from Etna, are identified in the chronology established using tephra layers and sapropels (organic-rich sediments). From Köng et al. (2016).*

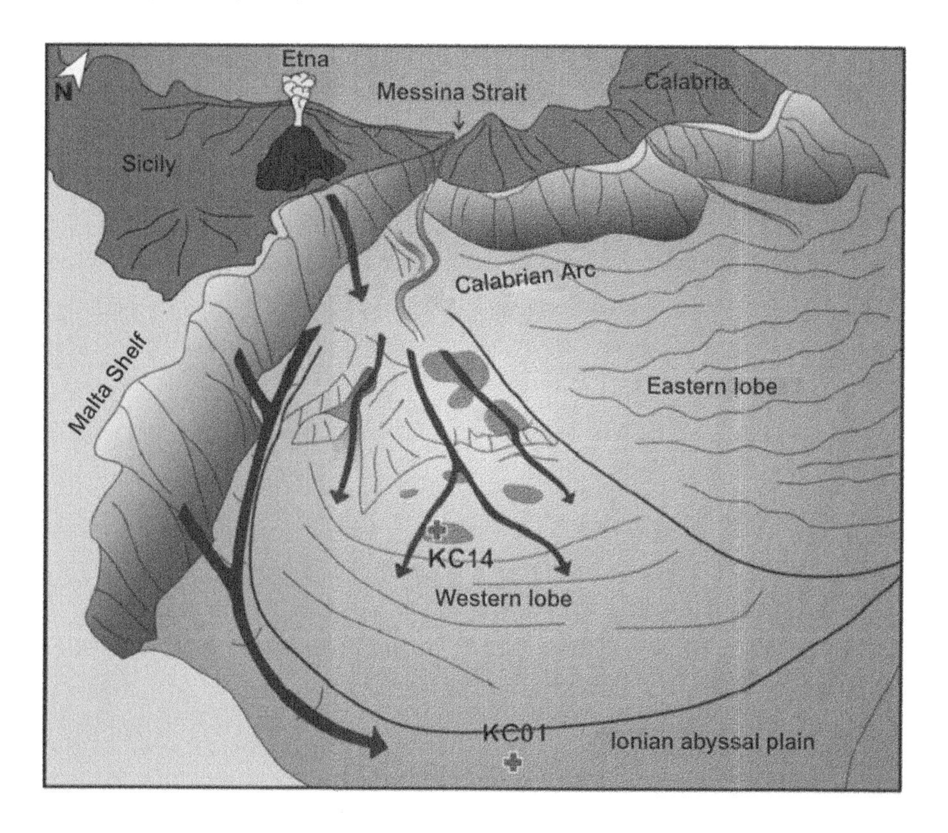

Figure 1.31. *Reconstruction of the main underwater gravity flows on the Calabrian arc (from Köng et al. (2016))*

1.10.2. *The recording of large eruptions in ice cores*

The use of polar ice cores as an archive to determine Earth's past climate is well known (Palais et al. 1987; NGICP 2004; Jouzel et al. 2007).

Ice cores, taken from the polar ice caps (Antarctica and Greenland), or from high altitude glaciers, are archives of the past climate, and also of the major internal and external events that have affected our planet or, for the recent period, of the human-induced industrial upheavals. Tephrochronology in ice cores is, for instance, a method used to study Icelandic volcanoes (Moles et al. 2019). Year after year, the accumulation of snowfall on Icelandic glaciers and ice caps traps ambient air, pollens and other particles in successive layers allowing their dating and the constitution of this frozen chronological archive. The oldest continuous ice cores can be dated back 123,000 years in Greenland and 800,000 years in Antarctica. The nature of the particles trapped in these cores varies according to the area from which they were taken.

Tephra ejected by eruptions powerful enough to send fragments and aerosols around the world are preserved in the cores, allowing them to be dated (Oppenheimer 2003; Davies et al. 2010; Narcisi et al. 2010). Thus, records of eruptions such as those of Vesuvius in 79 CE, Huaynaputina in 1600, Tambora in 1815, Krakatau in 1883 or El Chichon in 1982 have been identified in ice cores (see Figure 1.32) (Zielinski et al. 1997). The discovery of glassy particles associated with a particularly high volcanic sulfate aerosol record, found in Greenland and Antarctic ice cores, is also at the origin of the discovery of the Samalas eruption in 1257 (see section 1.1.1). The profound climatic changes from 43 and 42 BCE, which were among the coldest years in the Northern Hemisphere in recent millennia and marked the beginning of one of the coldest decades, are attributed to an eruption of Okmok volcano in Alaska (McConnell et al. 2020). These authors suggest that the climatic effects of this eruption may have induced the fall of the Roman Republic and led to the rise of the Roman Empire. In addition to identifying past eruptions, the work done on the cores can also provide complementary data on current major eruptions, as for the 1991 Pinatubo eruption with the estimate of the sulfur dioxide (SO_2) flux from ice core measurements (Cole-Dai et al. 1997). These approaches provide an opportunity to quantify the role of volcanism in ongoing climate change and to contribute to the study of the long-term relationship between eruptions, particularly very large eruptions, and climate variability.

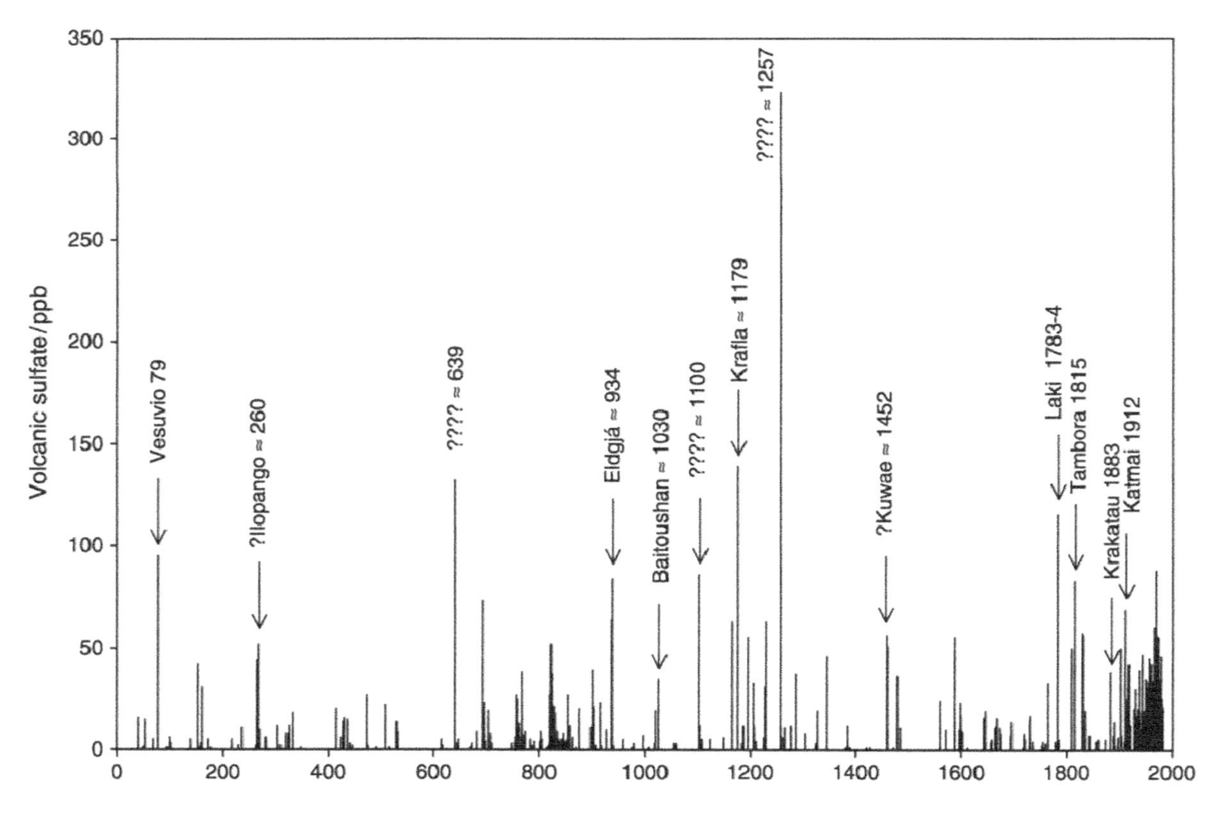

Figure 1.32. *Volcanic sulfate record in the GISP2 core (Greenland) for the first 20 centuries of our era*

COMMENT ON FIGURE 1.32.– *While many eruptions can be traced, several anomalies, including that of 1257, were not yet assigned to a specific volcano (see section 1.1.1) (after Oppenheimer 2003, origin of data and methods are presented in the publication).*

1.11. From the knowledge of a volcano's past to the identification of an operational monitoring strategy and the assessment of volcanic risks

We have thus seen various aspects of the construction of the geological history of volcanoes. By definition, this is a largely multidisciplinary task, as the approaches are so varied. This work must serve several key objectives of volcano monitoring:

– to bring a basic knowledge on the magmatic sources, types of magmas, their possible evolution and residence times, determining parameters to evaluate the rheological properties of magmas, their richness in gas;

– to establish the structural framework of the eruptions, the modalities of the magmatic injection, the existence of destabilization of the volcano flanks in the past, in connection or not with the eruptions;

– to evaluate the variability of eruptive regimes and the diversity of eruption styles, to establish a spatial and temporal mapping of hazards;

– to appreciate chronological variability and timescales.

All this knowledge allows us to establish strategies for geophysical and geochemical monitoring of a volcano, to establish scenarios for future eruptions, necessary for the constitution of risk management plans by the authorities and to make a diagnosis of the state of the volcano before and during an eruption.

When in 1977, following the eruption that partially destroyed the town of Piton Sainte-Rose, the decision was made to establish a volcanological observatory on Reunion Island (see Figure 1.33) to carry out operational monitoring of Piton de la Fournaise, geological knowledge of this volcano was limited. The chronology of the eruptions was partially known, in particular thanks to the work of Lacroix (1936, 1938) and to various reports published during the 1960s and 1970s. However, the diversity of eruptions, their spatial distribution, the sequence of events and their dynamic and petrological characteristics, necessary for the reflection that led to the setting up of the first observation networks, were the subject of work initiated to clarify the history of this volcano. From this work, and from the experience acquired by volcanologists at Kilauea in Hawaii, a volcano with similarities to Piton de la Fournaise, the first components of the observation and measurement strategy for monitoring were forged (Kornprobst et al. 1979; Bachèlery 1981; Chevallier et al. 1981; Chevallier and Bachèlery 1981; Bachèlery et al. 1982; Lénat and Aubert 1982; Bachèlery and Montaggioni 1983; Lénat and Bachèlery 1988). Thus, the conditions of lateral migration of magma toward short rift-zones with a fan-like morphology, the feeding of these eruptions by small magmatic reservoirs located at shallow depths under the summit of the volcano and the existence of recurrent explosive phases in the history of the volcano could be determined, offering the first pieces of a volcanic risk assessment (Villeneuve and Bachèlery 2006).

Figure 1.33. *The Piton de la Fournaise volcano observatory in the 1980s (photo: Jean-François Lénat)*

Today, probabilistic volcanic hazard assessment is becoming increasingly important (Marzocchi et al. 2004; Hincks et al. 2014; Connor et al. 2015). Based on the knowledge of the geological history of a volcano, it enables us to establish a probability of occurrence of the different types of eruptions, thus allowing a more quantitative anticipation of eruptions and facilitating the dialog with national or regional authorities in charge of civil security. Probabilistic risk maps for the various types of hazards identified for a volcano are now an important communication tool for raising awareness of volcanic risk (see Chapter 3).

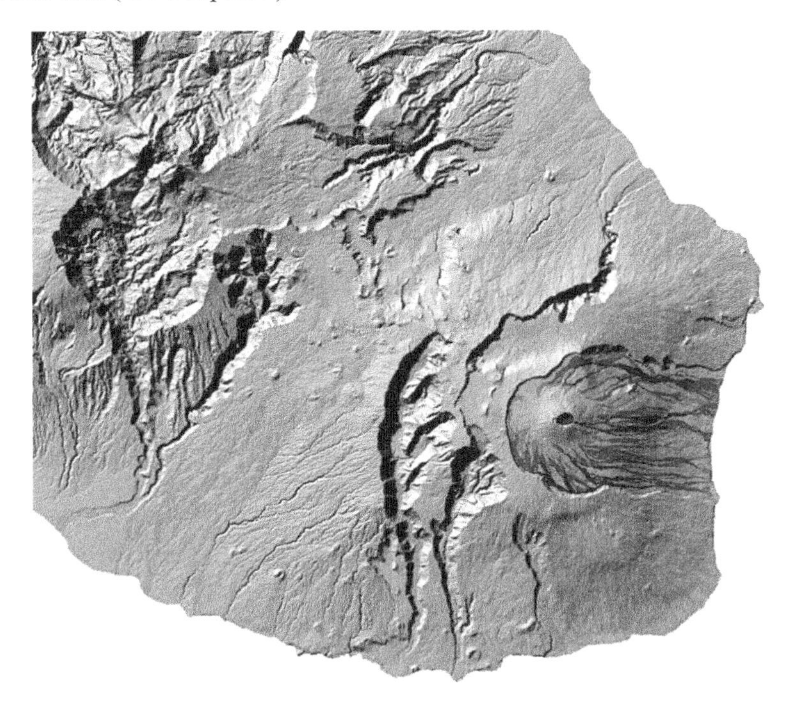

Figure 1.34. *Vulnerability map, derived from numerical simulations showing the probability of an eruptive fissure opening and the path of the associated flows, according to the type, location and frequency of past eruptions. For a color version of this figure, see www.iste.co.uk/lenat/hazards.zip*

COMMENT ON FIGURE 1.34.– *This type of document assesses the vulnerability of an area to lava flow invasion (from red: very high to green: low) (from Nave et al. 2016 modified).*

We began this chapter by emphasizing the importance of hazard mapping derived from the work of Crandell and Mullineaux (1978) at Mount St.

Helens, and inferred from their work on the eruptive history of that volcano. Such an approach remains more relevant than ever, with real improvements in our knowledge, in the quality of available data, in survey and imagery techniques offering high spatial and temporal resolution, in our understanding of the physics of volcanic processes, and in the computer processing of data. Simulations and models, which consider a wide range of factors, criteria and indicators to assess the probability of a particular eruptive scenario occurring, have developed in a few decades and are becoming increasingly important, also for operational monitoring purposes (see Figure 1.34). In parallel, the evolution of statistical models allows for better identification of uncertainties inherent to volcanic hazards, and a better consideration of diverse data sets, including subjective data through expert elicitation, in decision trees of volcanic events through Bayesian statistical methods (Neri et al. 2008; Hincks et al. 2014). The goal is to provide quantitative, probabilistic estimates of the occurrence and magnitude of potential volcanic events.

1.12. Conclusion

The geological history of a volcano is the knowledge base on which hazard identification and monitoring of a volcano's activity is founded. Any forecasting requires a good knowledge of the eruptive past of the volcano in order to be able to make a valid assessment of future eruptive behavior or the course of an ongoing eruption.

The acquisition of this knowledge implies approaching the history of the volcano at different temporal scales, using various types of archives. The historical activity (a few tens to hundreds of years) can be determined by the analysis of ancient works, narratives, descriptions, illustrations and archeological sites, and by the analysis of deposits left by the most recent eruptions, or of the vegetation covering them. Longer timescales (a few hundred to tens of thousands of years) are accessible from older sequences of deposits and geological structures resulting from the volcano's activity, and also by the study of more distant archives such as the record of eruptions in marine and lacustrine sediments, or by ice. Finally, the understanding of the structural and petrological context, and of the regional setting, complements this knowledge with data whose evolution is slower.

The work necessary to acquire this knowledge base must constitute the core of research on volcanoes, in addition to the geophysical and geochemical approaches implemented by volcanological observatories and research laboratories. Forecasting volcanic eruptions requires a good knowledge of the eruptive patterns specific to each volcano. This implies that each volcano be the subject of geological and volcanological studies, at the heart of which must be the determination of past eruptive dynamisms and regimes and the chronology of eruptions, in addition to geophysical, geochemical and magma monitoring (Agrinier et al. 2019).

1.13. References

Agrinier, P., Bachèlery, P., Bernard, P., Delouis, B., Deverchère, J., Grasso, J.-R. et al. (2019). *Quand la Terre tremble. Séismes et éruptions volcaniques et glissements de terrain en France*. CNRS Editions, Paris.

Ah-Peng, C., Chuah-Petiot, M., Descamps-Julien, B., Bardat, J., Stamenoff, P., Strasberg, D. (2007). Bryophyte diversity and distribution along an altitudinal gradient on a lava flow in La Réunion. *Diversity and Distributions*, 13(5), 654–662.

Albert, P.G., Hardiman, M., Keller, J., Smith, V.C., Bourne, A.J., Wulf, S. et al. (2015). Revisiting the Y-3 tephrostratigraphic marker: A new diagnostic glass geochemistry, age estimate, and details on its climatostratigraphical context. *Quat. Sci. Rev.*, 118, 105–121.

Albert, P.G., Tomlinson, E.L., Smith, V.C., Di Traglia, F., Pistolesi, M., Morris, A. et al. (2017). Glass geochemistry of pyroclastic deposits from the Aeolian Islands in the last 50 ka: A proximal database for tephrochronology. *J. Volcanol. Geotherm. Res.*, 336, 81–107.

Albert, S., Flores, O., Michon, L., Strasberg, D. (2020). Dating young (< 1 000 yr) lava flow eruptions of Piton de la Fournaise volcano from size distribution of long-lived pioneer trees. *Journal of Volcanology and Geothermal Research*, 401, 106974.

Allibert, C. (2015). L'archipel des Comores et son histoire ancienne. Essai de mise en perspective des chroniques, de la tradition orale et des typologies de céramiques locales et d'importation. *Afriques. Débats, méthodes et terrains d'histoire*. doi.org/10.4000/afriques.1721.

Anderson, K.R., Johanson, I.A., Patrick, M.R., Gu, M., Segall, P., Poland, M.P. et al. (2019). Magma reservoir failure and the onset of caldera collapse at Kīlauea Volcano in 2018. *Science*, 366(6470). doi.org/10.1126/science.aaz1822.

Andronico, D. and Cioni, R. (2002). Contrasting styles of Mount Visuvius activity in the period between the Avellino and Pompeii Plinian eruptions, and some implications for assessment of future hazards. *Bulletin of Volcanology*, 64(6), 372–391.

Arrighi, S., Principe, C., Rosi, M. (2001). Violent strombolian and subplinian eruptions at Vesuvius during post-1631 activity. *Bulletin of Volcanology*, 63(2–3), 126–150.

Atkinson, I. (1971). Successional trends in the coastal and lowland forests of Mauna Loa and Kilauea Volcano. *Hawaii: Pacific Science*, 24, 387–400.

Avvisati, G., Bellucci Sessa, E., Colucci, O., Marfè, B., Marotta, E., Nave, R. et al. (2019). Perception of risk for natural hazards in Campania Region (Southern Italy). *International Journal of Disaster Risk Reduction*, 40, 101164.

Babonneau, N., Villeneuve, N., Mazuel, A., Bachèlery, P. (2016). Erosion and volcaniclastic sedimentation at Piton de la Fournaise: From source to deep marine environment. In *Active Volcanoes of the World*, Bachélery, P., Lénat, J.-F., Di Muro, A., Michon, L. (eds). Springer, Berlin.

Bachèlery, P. (1981). Le Piton de la Fournaise (île de la Réunion). Étude volcanologique, structurale et pétrologique. PhD Thesis, Université de Clermont-Ferrand, Clermont-Ferrand.

Bachèlery, P. and Montaggioni, L.F. (1983). Morphostructure du flanc oriental du volcan de la Fournaise, île de la Réunion (Océan indien). *Comptes-Rendus Des Séances de l'Académie Des Sciences. Série 2, Mécanique-Physique, Chimie, Sciences de l'Univers, Sciences de La Terre*, 297(1), 81–84.

Bachèlery, P., Blum, P.A., Cheminée, J.-L., Chevallier, L., Gaulon, R., Girardin, N. et al. (1982). Eruption at Le Piton de la Fournaise volcano on 3 February 1981. *Nature*, 297(5865), 395–397.

Bachèlery, P., Morin, J., Villeneuve, N., Soulé, H., Nassor, H., Ali, A.R. (2016). Structure and eruptive history of Karthala Volcano. In *Active Volcanoes of the Southwest Indian Ocean: Piton de la Fournaise and Karthala, Active Volcanoes of the World*, Bachèlery, P., Lénat, J.-F., Di Muro, A., Michon, L. (eds). Springer, Berlin.

Barberi, F. and Carapezza, M.L. (2019). Explosive volcanoes in the mediterranean area: Hazards from future eruptions at Vesuvius (Italy) and Santorini (Greece). *Annals of Geophysics*, 62(1), 1–14.

Barberi, F., Rosi, M., Sodi, A. (1993). Volcanic hazard assessment at Stromboli based on review of historical data. *Acta Vulcanologica*, 3, 173–187.

Bazin, L., Lemieux-Dudon, B., Siani, G., Govin, A., Landais, A., Genty, D. et al. (2019). Construction of a tephra-based multi-archive coherent chronological framework for the last deglaciation in the Mediterranean region. *Quaternary Science Reviews*, 216, 47–57.

Behncke, B., Calvari, S., Giammanco, S., Neri, M., Pinkerton, H. (2008). Pyroclastic density currents resulting from the interaction of basaltic magma with hydrothermally altered rock: An example from the 2006 summit eruptions of Mount Etna, Italy. *Bulletin of Volcanology*, 70(10), 1249–1268.

Bertrand, S., Castiaux, J., Juvigné, E. (2008). Tephrostratigraphy of the late glacial and Holocene sediments of Puyehue Lake (Southern Volcanic Zone, Chile, 40°S). *Quaternary Research*, 70(3), 343–357.

Bory de Saint-Vincent, J.B.G.M. (1804). *Voyage dans les quatre principales îles des mers d'Afrique*, volume 3. Jeanne Laffitte Éditions, Marseille.

Boudon, G., Semet, M.P., Vincent, M.P. (1988). The 3100 and 11 500 BP flank failure/blast eruptions of La Grande Découverte (La Soufrière) volcano, Guadeloupe. *FWI Abstracts West Indies Explosive Volcanism Workshop*. Guadeloupe.

Boudon, G., Le Friant, A., Komorowski, J.C., Deplus, C., Semet, M.P. (2007). Volcano flank instability in the Lesser Antilles Arc: Diversity of scale, processes, and temporal recurrence. *Journal of Geophysical Research: Solid Earth*, 112(8). doi.org/10.1029/2006BJ004674.

Boudon, G., Villemant, B., Le Friant, A., Paterne, M., Cortijo, E. (2013). Role of large flank-collapse events on magma evolution of volcanoes. Insights from the Lesser Antilles Arc. *Journal of Volcanology and Geothermal Research*, 263, 224–237.

Bouma, A.H. (1962). *Sedimentology of Some Flysch Deposits; A Graphic Approach to Facies Interpretation*. Elsevier, Amsterdam.

Branca, S. and Abate, T. (2019). Current knowledge of Etna's flank eruptions (Italy) occurring over the past 2500 years. From the iconographies of the XVII century to modern geological cartography. *Journal of Volcanology and Geothermal Research*, 385, 159–178.

Branca, S. and Del Carlo, P. (2004). Eruptions of Mt. Etna during the past 3,200 years: A revised compilation integrating the historical and stratigraphic records. *Geophysical Monograph Series*, 143, 1–27.

Çağatay, M.N., Wulf, S., Sancar, Ü., Özmaral, A., Vidal, L., Henry, P. et al. (2015). The tephra record from the sea of marmara for the last ca. 70 ka and its palaeoceanographic implications. *Marine Geology*, 361, 96–110.

Calabrò, L., Harris, A.J.L., Thouret, J.C. (2020). Media views of the Stromboli 2002–2003 eruption and evacuation: A content analysis to understand framing of risk communication during a volcanic crisis. *Journal of Applied Volcanology*. doi.org/10.1186/s13617-020-00094-0.

Calvari, S., Cannavò, F., Bonaccorso, A., Spampinato, L., Pellegrino, A.G. (2018). Paroxysmal explosions, lava fountains and ash plumes at Etna volcano: Eruptive processes and hazard implications. *Frontiers in Earth Science*, 6(107). doi.org/10.3389/feart.2018.00107.

Carey, S.N., and Schneider, J.L. (2011). Volcaniclastic processes and deposits in the deep-sea. In *Developments in Sedimentology : Deep-Sea Sediments 63*, HüNeke, H. and Mulder, T. (eds). Elsevier, Amsterdam.

Carr, M.H. and Greeley, R. (1980). *Volcanic Features of Hawaii: A Basis for Comparison with Mars. Vol. 403*. National Aeronautics and Space Administration, Washington, D.C.

Carracedo, J.C., Day, S.J., Guillou, H., Pérez Torrado, F.J. (1999). Giant quaternary landslides in the evolution of La Palma and El Hierro, Canary Islands. *Journal of Volcanology and Geothermal Research*, 94(1–4), 169–190.

Cashman, K.V. and Giordano, G. (2008). Volcanoes and human history. *Journal of Volcanology and Geothermal Research*, 176(3), 325–329.

Cassidy, M., Watt, S.F.L., Palmer, M.R., Trofimovs, J., Symons, W., Maclachlan, S.E., Stinton, A.J. (2014). Construction of volcanic records from marine sediment cores: A review and case study (Montserrat, West Indies). *Earth-Science Reviews*, 138, 137–155.

Cassidy, M., Manga, M., Cashman, K., Bachmann, O. (2018). Controls on explosive-effusive volcanic eruption styles. *Nature Communications*, 9(1), 1–16.

Chevallier, L. and Bachèlery, P. (1981). Évolution structurale du volcan actif du Piton de la Fournaise, Île de la Réunion – Océan indien occidental. *Bulletin Volcanologique*, 44(4), 723–741.

Chevallier, L., Lalanne, F., Bachèlery, P., Vincent, P. (1981). L'éruption du mois de février 1981 au Piton de la Fournaise (Île de la Réunion – Océan indien). Phénomènologie et remarques structurales. *C.R. Acad. Sciences Paris*, 293, 187–190.

Cioni, R., Longo, A., Macedonio, G., Santacroce, R., Sbrana, A., Sulpizio, R., Andronico, D. (2003). Assessing pyroclastic fall hazard through field data and numerical simulations: Example from Vesuvius. *Journal of Geophysical Research: Solid Earth*, 108(B2). doi.org/10.1029/2001JB000642.

Cole-Dai, J., Mosley-Thompson, E., Thompson, L.G. (1997). Quantifying the Pinatubo volcanic signal in south polar snow. *Geophysical Research Letters*, 24(21), 2679–2682.

Connor, C., Bebbington, M., Marzocchi, W. (2015). Chapter 51 – Probabilistic volcanic hazard assessment. In *The Encyclopedia of Volcanoes*. doi.org/10.1016/B978-0-12-385938-9.00051-1.

Crandell, D.R. and Mullineaux, D.R. (1978). Potential hazards from future eruptions of Mount St. Helens Volcano, Washington. *USGS Bulletin*, 1383c, 1–26.

Crandell, D.R., Mullineaux, D.R., Rubin, M. (1975). Mount St. Helens volcano: Recent and future behaviour. *Science*, 187(4175), 438–441.

Crandell, D.R., Booth, B., Kusumadinata, K., Shimozuru, D., Walker, G.P.L., Westercamp, D. (1984). *Source Book for Volcanic Hazards Zonation*. Unesco Press, Paris.

Davies, S.M., Wastegård, S., Abbott, P.M., Barbante, C., Bigler, M., Johnsen, S.J. et al. (2010). Tracing volcanic events in the NGRIP ice-core and synchronising North Atlantic marine records during the last glacial period. *Earth and Planetary Science Letters*, 294(1–2), 69–79.

Day, S.J., Heleno Da Silva, S.I.N., Fonseca, J.F.B.D. (1999). A past giant lateral collapse and present-day flank instability of Fogo, Cape Verde Islands. *Journal of Volcanology and Geothermal Research*, 94(1–4), 191–218.

Deplus, C., Le Friant, A., Boudon, G., Komorowski, J.C., Villemant, B., Harford, C. et al. (2001). Submarine evidence for large-scale debris avalanches in the Lesser Antilles Arc. *Earth and Planetary Science Letters*, 192(2), 145–157.

Druitt, T.H., Mellors, R.A., Pyle, D.M., Sparks, R.S.J. (1989). Explosive volcanism on Santorini, Greece. *Geological Magazine*, 126(2), 95–126.

Druitt, T.H., Costa, F., Deloule, E., Dungan, M., Scaillet, B. (2012). Decadal to monthly timescales of magma transfer and reservoir growth at a caldera volcano. *Nature*, 482, 77–80.

Fearnley, C.J., Bird, D.K., Haynes, K., Mcguire, W.J., Jolly, G. (2018). *Advances in Volcanology: Observing the Volcano World*. Springer Verlag, Berlin.

Fretzdorff, S., Paterne, M., Stoffers, P., Ivanova, E. (2000). Explosive activity of the Reunion Island volcanoes through the past 260,000 years as recorded in deep-sea sediments. *Bulletin of Volcanology*, 62(4–5), 266–277.

Garcia, M.O. and Hull, D.M. (1994). Turbidites from giant Hawaiian landslides: Results from Ocean Drilling Program Site 842. *Geology*, 22(2), 159–162.

Gehrels, M.J., Lowe, D.J., Hazell, Z.J., Newnham, R.M. (2006). A continuous 5300-yr Holocene cryptotephrostratigraphic record from northern New Zealand and implications for tephrochronology and volcanic hazard assessment. *The Holocene*, 16, 173–187.

Germanaz, C. (2005). Du pont des navires au bord des cratères : regards croisés sur le Piton de la Fournaise (1653–1964). Itinéraires iconographiques et essai d'iconologie du volcan actif de La Réunion. PhD Thesis, Université Sorbonne-Panthéon, Paris.

Germanaz, C. (2016). Les dessous du dessus du volcan. Représenter l'évolution des cratères sommitaux de la Fournaise (1750–2007). In *Colloque Images & Savoirs*, Saint-Denis de La Réunion.

Geshi, N., Shimano, T., Chiba, T., Nakada, S. (2002). Caldera collapse during the 2000 eruption of Miyakejima Volcano, Japan. *Bulletin of Volcanology*, 64(1), 55–68.

Gorshkov, G.S. (1959). Gigantic eruption of the volcano Bezymianny. *Bulletin Volcanologique*, 20(1), 77–109.

Gudmundsdóttir, E.R., Eiríksson, J., Larsen, G. (2011). Identification and definition of primary and reworked tephra in Late Glacial and Holocene marine shelf sediments off North Iceland. *Journal of Quaternary Science*, 26(6), 589–602.

Gudmundsson, M.T., Jónsdóttir, K., Hooper, A., Holohan, E.P., Halldórsson, S.A., Ófeigsson, B.G., Einarsson, P. (2016). Gradual caldera collapse at Bárdarbunga volcano, Iceland, regulated by lateral magma outflow. *Science*, 353(6296), aaf8988.

Guillet, S., Corona, C., Stoffel, M., Khodri, M., Lavigne, F., Ortega, P. et al. (2017). Climate response to the Samalas volcanic eruption in 1257 revealed by proxy records. *Nature Geoscience*, 10(2), 123–128.

Gurioli, L., Sulpizio, R., Cioni, R., Sbrana, A., Santacroce, R., Luperini, W., Andronico, D. (2010). Pyroclastic flow hazard assessment at Somma-Vesuvius based on the geological record. *Bulletin of Volcanology*, 72(9), 1021–1038.

Hamann, Y., Wulf, S., Ersoy, O., Ehrmann, W., Aydar, E., Schmiedl, G. (2010). First evidence of a distal early Holocene ash layer in Eastern Mediterranean deep-sea sediments derived from the Anatolian volcanic province. *Quaternary Research*, 73(3), 497–506.

Hamilton, W. (1768). I. An account of the eruption of Mount Vesuvius, in 1767: In a letter to the Earl of Morton, President of the Royal Society, from the Honourable William Hamilton, his majesty's envoy extraordinary at Naples. *Philosophical Transactions of the Royal Society of London*, 58, 1–14.

Hincks, T.K., Komorowski, J.C., Sparks, S.R., Aspinall, W.P. (2014). Retrospective analysis of uncertain eruption precursors at La Soufrière volcano, Guadeloupe, 1975–77: Volcanic hazard assessment using a Bayesian Belief Network approach. *Journal of Applied Volcanology*, 3(1). doi.org/10.1186/2191-5040-3-3.

Holcomb, R.T. and Searle, R.C. (1991). Large landslides from oceanic volcanoes. *Marine Geotechnology*, 10(1–2), 19–32.

Hunt, J.E. and Jarvis, I. (2017). Prodigious submarine landslides during the inception and early growth of volcanic islands. *Nature Communications*, 8(1), 1–12.

Insinga, D.D., Tamburrino, S., Lirer, F., Vezzoli, L., Barra, M., De Lange, G.J. et al. (2014). Tephrochronology of the astronomically-tuned KC01B deep-sea core, Ionian Sea: Insights into the explosive activity of the Central Mediterranean area during the last 200 ka. *Quaternary Science Reviews*, 85, 63–84.

Jenkins, S.F., Barsotti, S., Hincks, T.K., Neri, A., Phillips, J.C., Sparks, R.S.J. et al. (2015). Rapid emergency assessment of ash and gas hazard for future eruptions at Santorini Volcano, Greece. *Journal of Applied Volcanology*, 4(1). doi.org/10.1186/s13617-015-0033-y.

Jerram, D. and Petford, N. (2012). *The Field Description of Igneous Rocks*, 2nd edition. Wiley-Blackwell, Hoboken.

Jousset, P.S., Pallister, J., Boichu, M., Buongiorno, M.F., Budisantoso, A. et al. (2012). The 2010 explosive eruption of Java's Merapi volcano: A "100-year" event. *Journal of Volcanology and Geothermal Research*, 241/242, 121–135.

Jouzel, J., Masson-Delmotte, V., Cattani, O., Dreyfus, G., Falourd, S., Hoffmann, G. et al. (2007). Orbital and millennial antarctic climate variability over the past 800,000 years. *Science*, 317(5839), 793–796.

Juvigné, E. and Miallier, D. (2016). Distribution, tephrostratigraphy and chronostratigraphy of the widespread eruptive products of Pavin Volcano. In *Lake Pavin: History, Geology, Biogeochemistry, and Sedimentology of a Deep Meromictic Maar Lake*. Springer, Berlin.

Keller, J., Ryan, W.B.F., Ninkovich, D., Altherr, R. (1978). Explosive volcanic activity in the Mediterranean over the past 200,000 yr as recorded in deep-sea sediments. *Bulletin of the Geological Society of America*, 89(4), 591–604.

Kokelaar, B.P. (2002). Setting, chronology and consequences of the eruption of Soufrière Hills Volcano, Montserrat (1995–1999). *Geological Society Memoir*, 21(1), 1–43.

Komorowski, J.-C., Boudon, G., Semet, M., Beauducel, F., Anténor-Habazac, C., Bazin, S., Hammouya, G. (2005). Guadeloupe. In *Volcanic Atlas of the Lesser Antilles, Seismic Research Unit*, Lindsay, J.M., Robertson, R.E.A., Shepherd, J.B., Ali, S. (eds). The University of the West Indies, Trinidad/Tobago.

Komorowski, J.-C., Legendre, Y., Caron, B., Boudon, G. (2008). Reconstruction and analysis of sub-plinian tephra dispersal during the 1530 A.D. Soufrière (Guadeloupe) eruption: Implications for scenario definition and hazards assessment. *Journal of Volcanology and Geothermal Research*, 178(3), 491–515.

Komorowski, J.-C., Morin, J., Jenkins, S., Kelman, I. (2018). Challenges of volcanic crises on small islands states. In *Observing the Volcano World*, Fearnley, C.J., Bird, D.K., Haynes, K., McGuire, W.J., Jolly, G. (eds). Springer Verlag, Berlin.

Köng, E., Zaragosi, S., Schneider, J.L., Garlan, T., Bachèlery, P., San Pedro, L. et al. (2016). Untangling the complex origin of turbidite activity on the Calabrian Arc (Ionian Sea) over the last 60 ka. *Marine Geology*, 373, 11–25.

Kornprobst, J., Boivin, P., Bachèlery, P. (1979). L'alimentation des éruptions récentes du Piton de la Fournaise (Île de la Réunion, Océan indien) : degré d'évolution et niveau de ségrégation des laves émises. *CR Acad. Sci. Paris*, 288, 1691–1694.

La Spina, G., Clarke, A.B., de' Michieli Vitturi, M., Burton, M., Allison, C.M., Roggensack, K., Alfano, F. (2019). Conduit dynamics of highly explosive basaltic eruptions: The 1085 CE Sunset Crater sub-Plinian events. *Journal of Volcanology and Geothermal Research*, 387, 106658.

Lacroix, A. (1936). Le volcan actif de l'Île de La Réunion. *Bulletin Volcanologique*. doi.org/10.1007/bf02596635.

Lacroix, A. (1938). *Le volcan actif de la Réunion (supplément) et celui de la Grande Comore*. Villars, Paris.

Lane, C.S., Andrič, M., Cullen, V.L., Blockley, S.P.E. (2011). The occurrence of distal Icelandic and Italian tephra in the Lateglacial of Lake Bled, Slovenia. *Quaternary Science Reviews*, 30(9–10), 1013–1018.

Lane, C.S., Brauer, A., Blockley, S.P.E., Dulski, P. (2013). Volcanic ash reveals time-transgressive abrupt climate change during the Younger Dryas. *Geology*, 41(12), 1251–1254.

Lavigne, F., Degeai, J.P., Komorowski, J.C., Guillet, S., Robert, V., Lahitte, P. et al. (2013). Source of the great A.D. 1257 mystery eruption unveiled, Samalas volcano, Rinjani Volcanic Complex, Indonesia. *Proceedings of the National Academy of Sciences of the United States of America*, 110, 16742–16747.

Le Friant, A., Deplus, C., Boudon, G., Sparks, R.S.J., Trofimovs, J., Talling, P. (2009). Submarine deposition of volcaniclastic material from the 1995–2005 eruptions of Soufrière Hills volcano, Montserrat. *Journal of the Geological Society*, 166(1), 171–182.

Le Friant, A., Lebas, E., Clément, V., Boudon, G., Deplus, C., De Voogd, B., Bachlery, P. (2011). A new model for the evolution of la Réunion volcanic complex from complete marine geophysical surveys. *Geophysical Research Letters*, 38(9). doi.org/10.1029/2011GL047489.

Le Friant, A., Ishizuka, O., Boudon, G., Palmer, M.R., Talling, P.J., Villemant, B. et al. (2015). Submarine record of volcanic island construction and collapse in the Lesser Antilles arc: First scientific drilling of submarine volcanic island landslides by IODP Expedition 340. *Geochemistry, Geophysics, Geosystems*, 16(2), 420–442.

Legendre, Y. (2012). Reconstruction fine de l'histoire éruptive et scénarii éruptifs à la Soufrière de Guadeloupe : vers un modèle intégré de fonctionnement du volcan, partie 1. PhD Thesis, Université Paris René Diderot, Paris.

Leicher, N., Giaccio, B., Zanchetta, G., Wagner, B., Francke, A., Palladino, D.M. et al. (2019). Central Mediterranean explosive volcanism and tephrochronology during the last 630 ka based on the sediment record from Lake Ohrid. *Quaternary Science Reviews*, 226, 106021.

Lénat, J.-F. and Aubert, M. (1982). Structure of Piton de la Fournaise volcano (La Reunion island, Indian ocean) from magnetic investigations. An illustration of the analysis of magnetic data in a volcanic area. *Journal of Volcanology and Geothermal Research*, 12(3/4). doi.org/10.1016/0377-0273(82)90035-X.

Lénat, J.-F. and Bachèlery, P. (1988). Dynamics of magma transfer at Piton de la Fournaise Volcano (Réunion Island, Indian Ocean). In *Modeling of Volcanic Processes*, Scarpa C.-Y. (ed.). Vieweg and Sohn, Brauschweig/Wiesbaden.

Lénat, J.-F., Vincent, P., Bachèlery, P. (1989). The off-shore continuation of an active basaltic volcano: Piton de la Fournaise (Réunion Island, Indian Ocean); structural and geomorphological interpretation from sea beam mapping. *Journal of Volcanology and Geothermal Research*. doi.org/10.1016/0377-0273(89)90003-6.

Lénat, J.-F., Bachèlery, P., Desmulier, F. (2001). Genèse du champ de lave de l'Enclos Fouque ; une éruption d'envergure exceptionnelle du Piton de la Fournaise (Réunion) au XVIIIe siècle. *Bulletin de La Société Géologique de France*, 172(2), 177–188.

Leone, F. and Lesales, T. (2006). Des cartes pour gérer le risque volcanique à la Martinique (Antilles françaises). *Revue internationale de géomatique*, 16, 341–358.

Lipman, P.W. and Mullineaux, D.R. (1981). The 1980 eruptions of Mount St. Helens, Washington. *US Geological Survey Professional Paper*, 1250. doi.org/10.1016/0377-0273(82)90073-7.

Lowe, D.J. (2011). Tephrochronology and its application: A review. *Quaternary Geochronology*, 6(2), 107–153.

Lucchi, F., Francalanci, L., De Astis, G., Tranne, C.A., Braschi, E., Klaver, M. (2019). Geological evidence for recurrent collapse-driven phreatomagmatic pyroclastic density currents in the Holocene activity of Stromboli volcano, Italy. *Journal of Volcanology and Geothermal Research*, 385, 81–102.

Luongo, G., Perrotta, A., Scarpati, C. (2003a). Impact of the AD 79 explosive eruption on Pompeii I. Relations amongst the depositional mechanisms of the pyroclastic products, the framework of the buildings and the associated destructive events. *Journal of Volcanology and Geothermal Research*, 126(3–4), 201–223.

Luongo, G., Perrotta, A., Scarpati, C., De Carolis, E., Patricelli, G., Ciarallo, A. (2003b). Impact of the AD 79 explosive eruption on Pompeii II. Causes of death of the inhabitants inferred by stratigraphic analysis and areal distribution of the human casualties. *Journal of Volcanology and Geothermal Research*, 126(3–4), 169–200.

MacDonald, G.A. (1965). Hawaiian calderas. *Pacific Science*, XIX, 320–334.

MacDonald, G.A. and Abbott, A.T. (1970). *Volcanoes in the Sea*. University of Hawaii Press, Honolulu.

Manville, V., Németh, K., Kano, K. (2009). Source to sink: A review of three decades of progress in the understanding of volcaniclastic processes, deposits, and hazards. *Sedimentary Geology*, 220(3–4), 136–161.

Martí, J. (2019). Las Cañadas caldera, Tenerife, Canary Islands: A review, or the end of a long volcanological controversy. *Earth-Science Reviews*, 196, 102889.

Marzocchi, W., Sandri, L., Gasparini, P., Newhall, C., Boschi, E. (2004). Quantifying probabilities of volcanic events: The example of volcanic hazard at Mount Vesuvius. *Journal of Geophysical Research: Solid Earth*, 109(11), 1–18.

Massin, F., Ferrazzini, V., Bachèlery, P., Nercessian, A., Duputel, Z., Staudacher, T. (2011). Structures and evolution of the plumbing system of Piton de la Fournaise volcano inferred from clustering of 2007 eruptive cycle seismicity. *Journal of Volcanology and Geothermal Research*, 202(1–2), 96–106.

Masson, D.G., Watts, A.B., Gee, M.J.R., Urgeles, R., Mitchell, N.C., Le Bas, T.P., Canals, M. (2002). Slope failures on the flanks of the western Canary Islands. *Earth-Science Reviews*, 57(1–2), 1–35.

Masson, D.G., Le Bas, T.P., Grevemeyer, I., Weinrebe, W. (2008). Flank collapse and large-scale landsliding in the Cape Verde Islands, off West Africa. *Geochemistry, Geophysics, Geosystems*, 9(7), 7015.

McConnell, J.R., Sigl, M., Plunkett, G., Burke, A., Kim, W.M., Raible, C.C. et al. (2020). Extreme climate after massive eruption of Alaska's Okmok volcano in 43 BCE and effects on the late Roman Republic and Ptolemaic Kingdom. *Proceedings of the National Academy of Sciences of the United States of America*, 117(27), 15443–15449.

Michon, L., Staudacher, T., Ferrazzini, V., Bachèlery, P., Marti, J. (2007). April 2007 collapse of Piton de la Fournaise: A new example of caldera formation. *Geophysical Research Letters*, 34(21), 21301.

Michon, L., Villeneuve, N., Catry, T., Merle, O. (2009). How summit calderas collapse on basaltic volcanoes: New insights from the April 2007 caldera collapse of Piton de la Fournaise volcano. *Journal of Volcanology and Geothermal Research*, 184(1–2), 138–151.

Michon, L., Massin, F., Famin, V., Ferrazzini, V., Roult, G. (2011). Basaltic calderas: Collapse dynamics, edifice deformation, and variations of magma withdrawal. *Journal of Geophysical Research: Solid Earth*, 116(3). doi.org/10.1029/2010JB007636.

Michon, L., Di Muro, A., Villeneuve, N., Saint-Marc, C., Fadda, P., Manta, F. (2013). Explosive activity of the summit cone of Piton de la Fournaise volcano (La Réunion island): A historical and geological review. *Journal of Volcanology and Geothermal Research*, 264, 117–133.

Mitchell, N.C. (2003). Susceptibility of mid-ocean ridge volcanic islands and seamounts to large-scale landsliding. *Journal of Geophysical Research*, 108(B8). doi.org/10.1029/2002jb001997.

Moles, J.D., McGarvie, D., Stevenson, J.A., Sherlock, S.C., Abbott, P.M., Jenner, F.E., Halton, A.M. (2019). Widespread tephra dispersal and ignimbrite emplacement from a subglacial volcano (Torfajökull, Iceland). *Geology*, 47. doi.org/10.1130/G46004.1.

Moore, J.G., Clague, D.A., Holcomb, R.T., Lipman, P.W., Normark, W.R., Torresan, M.E. (1989). Prodigious submarine landslides on the Hawaiian Ridge. *Journal of Geophysical Research*, 94(B12), 465–484.

Morandi, A., Di Muro, A., Principe, C., Michon, L., Leroi, G., Norelli, F., Bachèlery, P. (2016). Pre-historic (< 5 kiloyear) explosive activity at Piton de la Fournaise volcano. In *Active Volcanoes of the World*, Bachèlery, P., Lénat, J.-F., Di Muro, A., Michon, L. (eds). Springer, Berlin.

Moretti, R., Métrich, N., Arienzo, I., Di Renzo, V., Aiuppa, A., Allard, P. (2018). Degassing vs. eruptive styles at Mt. Etna volcano (Sicily, Italy). Part I: Volatile stocking, gas fluxing, and the shift from low-energy to highly explosive basaltic eruptions. *Chemical Geology*, 482, 1–17.

Morin, J., Bachèlery, P., Soulé, H., Nassor, H. (2016). Volcanic risk and crisis management on Grande Comore Island. In *Active Volcanoes of the World*, Bachèlery, P., Lénat, J.-F., Di Muro, A., Michon, L. (eds). Springer, Berlin.

Narcisi, B. (1996). Tephrochronology of a late quaternary lacustrine record from the Monticchio maar (Vulture volcano, southern Italy). *Quaternary Science Reviews*, 15(2–3), 155–165.

Narcisi, B., Petit, J.R., Chappellaz, J. (2010). A 70 ka record of explosive eruptions from the TALDICE ice core (Talos Dome, East Antarctic plateau). *Journal of Quaternary Science*, 25(6), 844–849.

Nave, R., Ricci, T., Pacilli, M.G. (2016). Perception of risk for volcanic hazard in Indian Ocean: La Réunion Island Case Study BT – Active volcanoes of the Southwest Indian Ocean: Piton de la Fournaise and Karthala. In *Active Volcanoes of the World*, Bachèlery, P., Lénat, J.-F., Di Muro, A., Michon, L. (eds). Springer, Berlin.

Neal, C.A., Brantley, S.R., Antolik, L., Babb, J.L., Burgess, M., Calles, K. et al. (2019). Volcanology: The 2018 rift eruption and summit collapse of Kilauea Volcano. *Science*, 363(6425), 367–374.

Neri, A., Aspinall, W.P., Cioni, R., Bertagnini, A., Baxter, P.J., Zuccaro, G. et al. (2008). Developing an event tree for probabilistic hazard and risk assessment at Vesuvius. *Journal of Volcanology and Geothermal Research*, 178(3), 397–415.

Newhall, C.G. and Self, S. (1982). The volcanic explosivity index (VEI): An estimate of explosive magnitude for historical volcanism. *Journal of Geophysical Research*, 87(C2), 1231–1238.

NGICP (2004). High-resolution record of Northern Hemisphere climate extending into the last interglacial period, by the North Greenland Ice Core Project members. *Nature*, 431(7005), 147–151.

Normark, W.R., Moore, J.G., Torresan, M.E. (1993). Giant volcano-related landslides and the development of the Hawaiian Islands. In *Submarine Landslides: Selected Studies in the US Exclusive Economic Zone*. US Geological Survey Bulletin, 2002, 184–196.

Oehler, J.-F., Lénat, J.-F., Labazuy, P. (2008). Growth and collapse of the Reunion Island volcanoes. *Bulletin of Volcanology*, 70(6), 717–742.

Oppenheimer, C. (2003). Ice core and palaeoclimatic evidence for the timing and nature of the great mid-13th century volcanic eruption. *International Journal of Climatology*, 23(4), 417–426.

Ort, M.H., Di Muro, A., Michon, L., Bachèlery, P. (2016). Explosive eruptions from the interaction of magmatic and hydrothermal systems during flank extension: The Bellecombe Tephra of Piton de La Fournaise (La Réunion Island). *Bulletin of Volcanology*, 78(1), 5.

Palais, J.M., Kyle, P.R., Mosley-Thompson, E., Thomas, E. (1987). Correlation of a 3,200 year old tephra in ice cores from Vostok and South Pole Stations, Antarctica. *Geophysical Research Letters*, 14(8), 804–807.

Pallister, J.S., Hoblitt, R.P., Crandell, D.R., Mullineaux, D.R. (1992). Mount St. Helens a decade after the 1980 eruptions: Magmatic models, chemical cycles, and a revised hazards assessment. *Bulletin of Volcanology*, 54(2), 126–146.

Paris, R., Bravo, J.J.C., González, M.E.M., Kelfoun, K., Nauret, F. (2017). Explosive eruption, flank collapse and megatsunami at Tenerife ca. 170 ka. *Nature Communications*, 8, 15246.

Paris, R., Ramalho, R.S., Madeira, J., Ávila, S., May, S.M., Rixhon, G. et al. (2018). Mega-tsunami conglomerates and flank collapses of ocean island volcanoes. *Marine Geology*, 395, 168–187.

Paterne, M., Guichard, F., Labeyrie, J. (1988). Explosive activity of the South Italian volcanoes during the past 80,000 years as determined by marine tephrochronology. *Journal of Volcanology and Geothermal Research*, 34(3–4), 153–172.

Paterne, M., Labeyrie, J., Guichard, F., Mazaud, A., Maitre, F. (1990). Fluctuations of the Campanian explosive volcanic activity (South Italy) during the past 190,000 years, as determined by marine tephrochronology. *Earth and Planetary Science Letters*, 98(2), 166–174.

Peltier, A., Bachèlery, P., Staudacher, T. (2009a). Magma transport and storage at Piton de La Fournaise (La Réunion) between 1972 and 2007: A review of geophysical and geochemical data. *Journal of Volcanology and Geothermal Research*, 184(1–2), 93–108.

Peltier, A., Staudacher, T., Bachèlery, P., Cayol, V. (2009b). Formation of the April 2007 caldera collapse at Piton de La Fournaise volcano: Insights from GPS data. *Journal of Volcanology and Geothermal Research*, 184(1–2), 152–163.

Peltier, A., Massin, F., Bachèlery, P., Finizola, A. (2012). Internal structure and building of basaltic shield volcanoes: The example of the Piton de La Fournaise terminal cone (La Réunion). *Bulletin of Volcanology*, 74(8), 1881–1897.

Perinotto, H., Schneider, J.L., Bachèlery, P., Le Bourdonnec, F.X., Famin, V., Michon, L. (2015). The extreme mobility of debris avalanches: A new model of transport mechanism. *Journal of Geophysical Research: Solid Earth*, 120(12), 8110–8119.

Piper, D.J.W. and Normark, W.R. (2009). Processes that initiate turbidity currents and their influence on turbidites: A marine geology perspective. *Journal of Sedimentary Research*, 79(6), 347–362.

Pistolesi, M., Bertagnini, A., Di Roberto, A., Ripepe, M., Rosi, M. (2020). Tsunami and tephra deposits record interactions between past eruptive activity and landslides at Stromboli volcano, Italy. *Geology*, 48(5), 436–440.

Prival, J.M., Thouret, J.C., Japura, S., Gurioli, L., Bonadonna, C., Mariño, J., Cueva, K. (2020). New insights into eruption source parameters of the 1600 CE Huaynaputina Plinian eruption, Peru. *Bulletin of Volcanology*, 82(1). doi.org/10.1007/s00445-019-1340-7.

Pyle, D.M., Ricketts, G.D., Margari, V., van Andel, T.H., Sinitsyn, A.A., Praslov, N.D., Lisitsyn, S. (2006). Wide dispersal and deposition of distal tephra during the Pleistocene "Campanian Ignimbrite/Y5" eruption, Italy. *Quaternary Science Reviews*, 25(21–22), 2713–2728.

Riel, B., Milillo, P., Simons, M., Lundgren, P., Kanamori, H., Samsonov, S. (2016). The collapse of Bárðarbunga caldera, Iceland. *Geophysical Journal International*, 202(1), 446–453.

Romero, J.E., Ramírez, V., Alam, M.A., Bustillos, J., Guevara, A., Urrutia, R. et al. (2020). Pyroclastic deposits and eruptive heterogeneity of Volcán Antuco (37°S; Southern Andes) during the Mid to Late Holocene (< 7.2 ka). *Journal of Volcanology and Geothermal Research*, 392, 106759.

Rowland, S.K. (2020). Goldschmidt virtual conference 2020 [Online]. Available at: https://www.higp.hawaii.edu/~scott/Fieldguide/Kilauea_fieldguide09.pdf.

Rubin, M. (1987). Hawaiian radiocarbon dates. *US Geol. Surv. Prof. Pap.*, 1350, 213–242.

Scandone, R. and Giacomelli, L. (2013). Volcanic precursors in light of eruption mechanisms at Vesuvius. *Annals of Geophysics*, 56(4), 4401–6461.

Scandone, R., Giacomelli, L., Gasparini, P. (1993). Mount Vesuvius: 2000 years of volcanological observations. *Journal of Volcanology and Geothermal Research*, 58(1–4), 5–25.

Scandone, R., Giacomelli, L., Speranza, F.F. (2008). Persistent activity and violent strombolian eruptions at Vesuvius between 1631 and 1944. *Journal of Volcanology and Geothermal Research*, 170(3–4), 167–180.

Scarpati, C., Perrotta, A., Martellone, A., Osanna, M. (2020). Pompeian hiatuses: New stratigraphic data highlight pauses in the course of the AD 79 eruption at Pompeii. *Geological Magazine*, 157(4), 695–700.

Schneider, J.-L., Le Ruyet, A., Chanier, F., Buret, C., Ferrière, J., Proust, J.-N., Rosseel, J.-B. (2001). Primary or secondary distal volcaniclastic turbidites: How to make the distinction? An example from the Miocene of New Zealand (Mahia Peninsula, North Island). *Sedimentary Geology*, 145(1), 1–22.

Sevink, J., Di Vito, M.A., van Leusen, P.M., Field, M.H. (2019). Distal effects of volcanic eruptions on pre-industrial societies. *Quaternary International*, 499, 129–134.

Sigl, M., Winstrup, M., McConnell, J.R., Welten, K.C., Plunkett, G., Ludlow, F. et al. (2015). Timing and climate forcing of volcanic eruptions for the past 2,500 years. *Nature*, 523(7562), 543–549.

Sigmundsson, F. (2019). Calderas collapse as magma flows into rifts. *Science*, 366(6470), 1200–1201.

Sigmundsson, F., Hooper, A., Hreinsdóttir, S., Vogfjörd, K.S., Ófeigsson, B.G., Heimisson, E.R. et al. (2014). Segmented lateral dyke growth in a rifting event at Bárðarbunga volcanic system, Iceland. *Nature*, 517(7533), 191–195.

Sigurdsson, H., Cashdollar, S., Sparks, S.R.J. (1982). The eruption of Vesuvius in AD 79: Reconstruction from historical and volcanological evidence. *American Journal of Archaeology*, 86(1), 39.

Sigurdsson, H., Carey, S., Cornell, W., Pescatore, T. (1985). The eruption of Vesuvius in AD 79. *National Geographic Research*, 1(3), 332–387.

Simkin, T. and Howard, K.A. (1970). Caldera collapse in the Galapagos Islands, 1968: The largest known collapse since 1912 followed a flank eruption and explosive volcanism within the caldera. *Science*, 169(3944), 429–437.

Sparks, R.S.J. (2003). Forecasting volcanic eruptions. *Earth and Planetary Science Letters*, 210(1–2), 1–15.

Sparks, R.S.J. and Young, S.R. (2002). The eruption of Soufrière Hills Volcano, Montserrat (1995–1999): Overview of scientific results. *Geological Society Memoir*, 21(1), 45–69.

Sparks, R.S.J., Barclay, J., Calder, E.S., Herd, R.A., Komorowski, J.C., Luckett, R., Norton, G.E., Ritchie, L.J., Voight, B., Woods, A.W. (2002). Generation of a debris avalanche and violent pyroclastic density current on 26 December (Boxing Day), 1997 at Soufriere Hills Volcano, Montserrat. *Geological Society Memoir*, 21(1), 409–434.

Staudacher, T., Ferrazzini, V., Peltier, A., Kowalski, P., Boissier, P., Catherine, P. et al. (2009). The April 2007 eruption and the Dolomieu crater collapse, two major events at Piton de la Fournaise (La Réunion Island, Indian Ocean). *Journal of Volcanology and Geothermal Research*, 184(1), 126–137.

Staudacher, T., Peltier, A., Ferrazzini, V., Di Muro, A., Boissier, P., Catherine, P. et al. (2016). Fifteen years of intense eruptive activity (1998–2013) at Piton de la Fournaise volcano: A review. In *Active Volcanoes of the World*, Bachèlery, P., Lénat, J.-F., Di Muro, A., Michon, L. (eds). Springer, Berlin.

Sulpizio, R., Bonasia, R., Dellino, P., Di Vito, M.A., La Volpe, L., Mele, D. et al. (2008). Discriminating the long distance dispersal of fine ash from sustained columns or near ground ash clouds: The example of the Pomici di Avellino eruption (Somma-Vesuvius, Italy). *Journal of Volcanology and Geothermal Research*, 177(1), 263–276.

Sulpizio, R., Van Welden, A., Caron, B., Zanchetta, G. (2010). The Holocene tephrostratigraphic record of Lake Shkodra (Albania and Montenegro). *Journal of Quaternary Science*, 25(5), 633–650.

Swanson, D.A. (2008). Hawaiian oral tradition describes 400 years of volcanic activity at Kīlauea. *Journal of Volcanology and Geothermal Research*, 176(3), 427–431.

Swanson, D.A., Rose, T.R., Fiske, R.S., McGeehin, J.P. (2012). Keanakāko'i Tephra produced by 300 years of explosive eruptions following collapse of Kīlauea's caldera in about 1500 CE. *Journal of Volcanology and Geothermal Research*, 215/216, 8–25.

Swanson, D.A., Weaver, S.J., Houghton, B.F. (2014). Reconstructing the deadly eruptive events of 1790 CE at Kīlauea Volcano, Hawaii. *Bulletin of the Geological Society of America*, 127(3–4), 503–515.

Tanguy, J.C., Bachèlery, P., LeGoff, M. (2011). Archeomagnetism of Piton de la Fournaise: Bearing on volcanic activity at La Réunion Island and geomagnetic secular variation in Southern Indian Ocean. *Earth and Planetary Science Letters*, 303(3–4), 361–368.

Thouret, J.C., Juvigné, E., Gourgaud, A., Boivin, P., Dávila, J. (2002). Reconstruction of the AD 1600 Huaynaputina eruption based on the correlation of geologic evidence with early Spanish chronicles. *Journal of Volcanology and Geothermal Research*, 115(3–4), 529–570.

Tilling, R.I., Topinka, L.J., Swanson, D.A. (1990). Eruptions of Mount St. Helens: Past, present and future. Report, U.S. Geological Survey, Reston, VA.

Trofimovs, J., Sparks, R.S.J., Talling, P.J. (2008). Anatomy of a submarine pyroclastic flow and associated turbidity current: July 2003 dome collapse, Soufrière Hills volcano, Montserrat, West Indies. *Sedimentology*, 55(3), 617–634.

Troll, V.R., Deegan, F.M., Jolis, E.M., Budd, D.A., Dahren, B., Schwarzkopf, L.M. (2015). Ancient oral tradition describes volcano-earthquake interaction at Merapi volcano, Indonesia. *Geografiska Annaler, Series A: Physical Geography*, 97(1), 137–166.

Ui, T., Takarada, S., Yoshimoto, M. (2000). Debris avalanches. In *Encyclopedia of Volcanoes*, Sigurdsson, H., Houghton, B., McNutt, S.R., Rymer, H., Stix, J. (eds). Academic Press, San Diego.

Van Wyk De Vries, B. and Davies, T. (2015). Landslides, debris avalanches, and volcanic gravitational deformation. In *Encyclopedia of Volcanoes*, Sigurdsson, H., Houghton, B., McNutt, S.R., Rymer, H., Stix, J. (eds). Academic Press, San Diego.

Van Wyk De Vries, B., Self, S., Francis, P.W., Keszthelyi, L. (2001). A gravitational spreading origin for the Socompa debris avalanche. *Journal of Volcanology and Geothermal Research*, 105(3), 225–247.

Vélain, C. (1878). *Description géologique de la presqu'île d'Aden, de l'île de La Réunion, des îles Saint-Paul et Amsterdam*. Hennuyer A., Paris.

Vidal, C.M., Komorowski, J.C., Métrich, N., Pratomo, I., Kartadinata, N., Prambada, O. et al. (2015). Dynamics of the major plinian eruption of Samalas in 1257 A.D. (Lombok, Indonesia). *Bulletin of Volcanology*, 77(9), 73.

Vidal, C.M., Métrich, N., Komorowski, J.C., Pratomo, I., Michel, A., Kartadinata, N. et al. (2016). The 1257 Samalas eruption (Lombok, Indonesia): The single greatest stratospheric gas release of the Common Era. *Scientific Reports*, 6, 34868.

Villeneuve, N. and Bachèlery, P. (2006). Revue de la typologie des éruptions au Piton de La Fournaise, processus et risques volcaniques associés. *Cybergeo*, 2006, 1–25.

Vincent, P.M., Bourdier, J.L., Boudon, G. (1989). The primitive volcano of Mount Pelée: Its construction and partial destruction by flank collapse. *Journal of Volcanology and Geothermal Research*, 38(1–2), 1–15.

Voight, B., Komorowski, J.C., Norton, G.E., Belousov, A.B., Belousova, M., Boudon, G. et al. (2002). The 26 December (Boxing Day) 1997 sector collapse and debris avalanche at Soufrière Hills Volcano, Montserrat. *Geological Society Memoir*, 21(1), 363–407.

Vougioukalakis, G., Sparks, R.S., Druitt, T., Pyle, D., Papazachos, C., Fytikas, M. (2017). Volcanic hazard assessment at Santorini Volcano: A review and a synthesis in the light of the 2011–2012 Santorini unrest. *Bulletin of the Geological Society of Greece*, 50(1), 274.

Watkins, N.D., Sparks, R.S.J., Sigurdsson, H., Huang, T.C., Federman, A., Carey, S., Ninkovich, D. (1978). Volume and extent of the Minoan tephra from Santorini Volcano: New evidence from deep-sea sediment cores. *Nature*, 271(5641), 122–126.

Zanchetta, G., Sulpizio, R., Roberts, N., Cioni, R., Eastwood, W.J., Siani, G. et al. (2011). Tephrostratigraphy, chronology and climatic events of the Mediterranean basin during the Holocene: An overview. *Holocene*, 21(1), 33–52.

Zielinski, G.A., Dibb, J.E., Yang, Q., Mayewski, P.A., Whitlow, S., Twickler, M.S., Germani, M.S. (1997). Assessment of the record of the 1982 El Chichón eruption as preserved in Greenland snow. *Journal of Geophysical Research Atmospheres*, 102(25), 30031–30045.

Zubia, M., De Clerck, O., Leliaert, F., Payri, C., Mattio, L., Vieira, C. et al. (2018). Diversity and assemblage structure of tropical marine flora on lava flows of different ages. *Aquatic Botany*, 144, 20–30.

2

Volcanic Hazards

Raphaël PARIS[1], Philipson BANI[1], Oryaëlle CHEVREL[1], Franck DONNADIEU[1], Julia EYCHENNE[1], Pierre-Jean GAUTHIER[1], Mathieu GOUHIER[1], David JESSOP[1,2], Karim KELFOUN[1], Séverine MOUNE[1,2], Olivier ROCHE[1] and Jean-Claude THOURET[1]

[1] *Laboratoire Magmas et Volcans, CNRS, IRD, OPGC, Université Clermont Auvergne, Clermont-Ferrand, France*

[2] *OVSG, Gourbeyre, Guadeloupe Institut de physique du globe de Paris, Université de Paris, Paris, France*

2.1. Introduction

The complexity of understanding volcanic risk is partly due to the fact that it is the result of different hazards, some of which are directly linked to the eruptive activity (gas, lava flows, pyroclastic flows and ash fallout) and others which are directly or indirectly induced by these hazards (debris avalanches, tsunamis, mudflows or lahars). There are therefore different eruptive styles, different types of eruptions and a multitude of possible eruptive scenarios. With the growth of the world's population, it is estimated today that the volcanic risk concerns about 800 million people (living within 100 km of an active or potentially active volcano) spread over 86 countries. Over the past four centuries, volcanoes have caused the deaths of 280,000 people, including 163,000 related to the five deadliest eruptions (Auker et al. 2013): Unzen 1792 (Japan), Tambora 1815 (Indonesia), Krakatau 1883 (Indonesia), Mont Pelée 1902 (Martinique) and Nevado del Ruiz 1985 (Colombia). The majority of eruptions are not directly lethal but severe damage has been caused by some eruptions (densely populated areas, poorly anticipated scenarios, cascading hazards, etc.). Volcanic crises can be spread out over time, over several months

Hazards and Monitoring of Volcanic Activity 1,
coordinated by Jean-François LÉNAT. © ISTE Ltd 2022.

or even years (Lanzarote eruption from 1730 to 1736; quasi-permanent eruption of Kilauea from 1983 to 2018). The impact of a volcanic eruption is also very different depending on whether you are in the proximal zone, on the flanks of the volcano (areas invaded by lava flows, pyroclastic flows and thick layers of ash), or in the distal zone (air traffic disruptions due to ash).

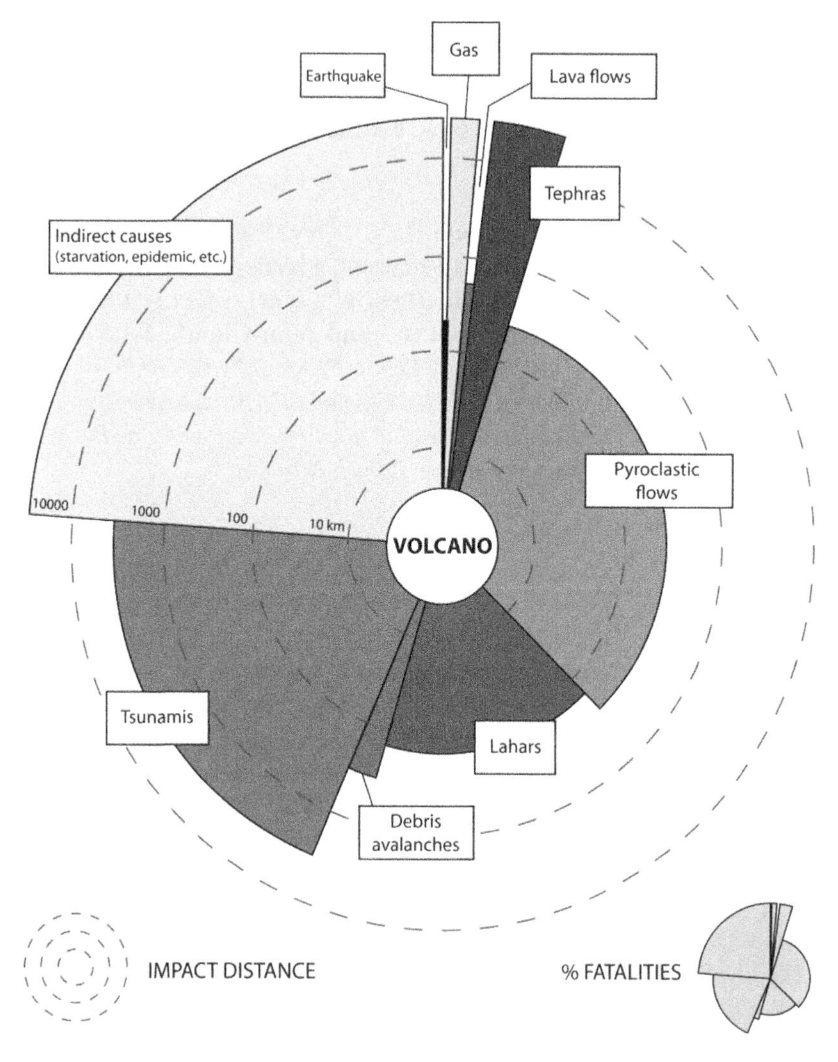

Figure 2.1. *Fatalities related to volcanic hazards (Auker et al. 2013) and their impact distance (R. Paris)*

The human toll (see Figure 2.1) of natural disasters is quickly estimated, but the economic toll can be spread over several weeks or even several years for the largest eruptions. In statistics, the economic impact of natural disasters is often reduced to an estimate of the overall cost. The 2011 Japanese earthquake and tsunami, for example, cost 210 billion dollars. Floods, storms and tsunamis are indeed particularly devastating, both in human and economic terms (impact often estimated at more than 10 billion dollars per occurrence). In comparison, most volcanic eruptions in recent decades (1980–2010) have cost less than 1 billion dollars, with the exception of the eruption of the Icelandic volcano Eyjafjallajökull in 2010 (at least 1.7 billion dollars) and Unzen in Japan in 1991–1993 (1.5 billion dollars). However, the deadliest disasters are not necessarily the most expensive on a global scale. The economic impact of a volcanic eruption can be spread over several months or even years. Eruptions such as Laki (Iceland, 1783) or Tambora (Indonesia, 1815) have had a regional and even global climatic and economic impact. Although difficult to estimate, the economic consequences of these historical eruptions were very important.

2.2. Eruptive hazards

2.2.1. *Earthquakes of magmatic and volcano-tectonic origin*

A wide range of seismic phenomena is associated with volcanic activity. Their monitoring and analysis are discussed extensively in Chapter 1, "Seismic Monitoring of Volcanoes and Eruption Forecasting", of Volume 2 of this series of books. The main sources of seismic signals are:

– magma transfers;

– hydrothermal activity;

– volcano-tectonic phenomena.

Most volcanic earthquakes are of low to moderate magnitude and, therefore, do not represent a hazard in most cases. However, they can be associated with significant deformation and surface fracture formation. The repetition of small earthquakes can create damage (boulders falling from unstable walls). Earthquakes of high magnitude, such as those associated with large-scale volcano-tectonic phenomena (flank slide, caldera collapse), have the same effects as tectonic earthquakes.

2.2.2. *Outgassing phenomena*

Understanding the hazard associated with volcanic gases means first of all understanding the physico-chemistry of the outgassing process. Initially dissolved at depth in the silicate liquids that form the magmas, many volatile molecular species will form an exsolved gas phase as the magmas rise to the surface. Thus, it is a liquid-gas biphasic medium, possibly accompanied by solid crystals that feed the volcanic edifices. However, at lower depths where the pressure is also lower, the bubbles grow and occupy a larger and larger volume. This increase in volume leads to a de facto decrease in the density of the magma, allowing it to rise faster and faster to the surface. For this reason, the outgassing of magmatic volatiles is often considered to drive eruptions (Sparks 1978; Papale and Polacci 1999; Shea 2017). Nevertheless, beyond this driving role, the outgassing process itself strongly influences the eruptive dynamics depending on whether the gas phase causes the fragmentation of the magma or not. It is simultaneously or independently released from the liquid phase. Strombolian dynamics where a small quantity of lava shreds is expelled from a crater by the surface bursting of metric gas bubbles, on the one hand, and Plinian dynamics where the complete fragmentation of the magma into a mixture of gas and fine ash, on the other hand, represent two extreme cases illustrating the importance of the physics of the outgassing process on the eruptive phenomenology. Volcanic hazard in the broadest sense is therefore strongly conditioned by magmatic outgassing itself. These aspects will be dealt with in the other sections of this chapter on hazard and will not be repeated here, where we will focus solely on the hazard associated with the gas itself. Similarly, the reader may refer to Chapter 1, "Volcanic Fluid Monitoring," in Volume 3 of this series of books for details of how magmatic outgassing manifests itself at the surface, from passive open-duct outgassing to high- and low-temperature hydrothermal events.

The hazard generated by volcanic gases is intrinsically linked to the chemical composition of the gas phase. This is mainly composed of six species (see Table 2.1) in varying proportions depending on the volcanoes and their geodynamic context, which together account for almost 100% of the gas: water vapor (H_2O), carbon dioxide (CO_2), sulfur species (SO_2 and H_2S) and halogenated species (HCl and HF). To these major compounds are added other minor gaseous species (e.g. rare gases such as He or Ar, CO, CH_4 and other alkanes, H_2 and nitrogenous compounds such as N_2 or NO_x) as well as a host of metallic, alkaline or alkaline-earth trace elements that form compounds, most often halides and sulfosalts, which are gaseous at magma temperature.

All these chemical species are obviously not equally dangerous in terms of environmental and societal impacts and generate a multi-faceted hazard. It is therefore necessary to distinguish between hazards on a global scale, which are likely to modify climatic and environmental conditions and disrupt ecosystems in the long term, and hazards on a local scale, which have more limited and often shorter-lived effects, but which are potentially dramatic for the populations exposed.

Geodynamic context	Mid-ocean ridges	Oceanic hot spot islands	Subduction zones	Subduction zones	Subduction zones
Nature of magmas	Basalt (MORB)	Basalt (OIB)	Arc basalt	Andesite	Dacite and rhyolite
Pre-eruptive abundance (in mass) of volatiles in deep magmas					
H_2O (%)	<0.4–0.5	0.2–1	0.2–6	>3	3–7
CO_2 (ppm)	50–400	2,000–6,500	25–2,500	10–1,200	<d.l.
S (ppm)	800–1,500	0–3,000	900–2,500	0–1,000	0–200
Cl (ppm)	20–50	20–90	500–2,000	100–1,500	600–2,700
F (ppm)	<100–600	<100–600	<500	<500	200–1,500
Molar concentrations (in %) of the main volatile species in volcanic gases					
H_2O	65–85	40–65	40–55	90–98	97–100
CO_2	5–25	20–45	20–30	1–5	0–2
SO_2	4–10	10–20	20–35	0–3	0–0.15
H_2S	0.5–1.5	0.05–0.25	0.15–0.25	0.1–0.9	0–0.35
HCI	<0.5	0.02–0.25	4–10	0.2–3.5	0.05–0.15
HF	<d.l.	<d.l.	0.5–2	0.0–0.08	0.01–0.03

Table 2.1. *Pre-eruptive abundance of volatiles in different magma types (from Wallace (2005) and Oppenheimer et al. (2014)) and related volcanic gas composition (from Symonds et al. (1994) and Oppenheimer et al. (2014)). The abbreviation <d.l. stands for "below detection limit"*

2.2.2.1. *Global climate effects: surface cooling and volcanic winters*

The long-term and large-scale effects of volcanic outgassing are always related to massive injections of gases into the atmosphere during particularly violent and/or long eruptions. Reactive gases (e.g. SO_2) are those that will have the strongest impact on the physico-chemistry of the atmosphere and Earth's radiation balance. Volcanic sulfur dioxide, injected into the atmosphere, reacts rapidly with hydroxyl radicals, oxygen and even ozone in the atmosphere to form sulfuric acid or sulfate aerosols. Several reactive pathways exist (Martin 2018), two of which are particularly important:

$$\begin{aligned} &SO_2 + OH \rightarrow HOSO_2 \\ &HOSO_2 + O_2 \rightarrow SO_3 + HO_2 \\ &SO_3 + H_2O \rightarrow H_2SO_4 \end{aligned} \qquad [2.1]$$

and:

$$SO_2{\bullet}OH\ (aq) + O_3 \rightarrow H^+ + SO_4^{2-} + O_2 \qquad [2.2]$$

Whether by simple oxidation or by ozone destruction, the injection of SO_2 into the atmosphere leads to the formation of a layer of sulfate aerosols. This layer reflects a significant proportion of the solar radiation and therefore generates abnormally low ground temperatures. Obviously, if these physico-chemical reactions occur in the upper atmosphere (stratosphere), the residence time of the aerosols is all the longer and therefore the thermal disturbance all the more marked (see Figure 2.2). This is known as a volcanic winter, a phenomenon first described after the eruption of Laki in Iceland in 1783 by Benjamin Franklin, who was on a diplomatic mission in Paris. He described "a persistent fog, essentially dry, on which the Sun had no effect of dissipation. There followed an early autumn when the land froze prematurely and a particularly cold and snowy winter such as France had never known before." Post hoc reconstructions, based on recent atmospheric models, suggest that the 122 megatons of SO_2 injected into the atmosphere generated a surface cooling of more than 3°C in Europe and North America until 1786, more than 3 years after the eruption (Zambri et al. 2019). Such a brutal cooling is, of course, heavy with consequences at the societal level with crop losses and therefore famines, increased poverty and diseases. Some authors thus consider that these "years without summer" following the eruption of Laki plunged the French peasantry into increased distress, an aggravating factor in the popular uprising that led to the French Revolution. There are many examples in recent history of these "years without summer" following large-scale volcanic eruptions, the term having been used initially for the year 1816, following the eruption of the Tambora volcano (Indonesia), in 1815. Again, a global effect of this eruption could be recorded with summer temperatures 3–5°C below seasonal normals in continental Europe and North America (about 1°C of this cooling being directly related to the stratospheric layer of sulfate aerosols), with, for example, snowfall on the east coast of the United States until June (Brönnimann and Krämer 2016). The electrostatic levitation of ash and aerosols more than 50 km aboveground in the ionosphere led to the disruption of the electrical balance of the atmosphere, resulting in heavy rainfall in Europe 2 months after the eruption, in June 1815, during Napoleon's defeat at Waterloo (Wheeler and Demarée 2005). More recently, the 1991 eruption of Pinatubo (Philippines) injected 20 megatons of SO_2 into the stratosphere, generating a global temperature decrease of about 0.4°C.

2.2.2.2. *Global climate effects: radiative forcing and temperature rise*

In contrast to SO_2, volcanic gases such as CO_2 or HCl will directly or indirectly generate a radiative forcing (commonly but incorrectly called "greenhouse effect") and thus contribute to global warming. The injection of HCl into the upper atmosphere causes the destruction of stratospheric ozone (O_3) according to the following reactions, after dissociation of HCl:

$$\begin{array}{l} Cl + O_3 \rightarrow ClO + O_2 \\ ClO + O \rightarrow Cl + O_2 \end{array} \qquad [2.3]$$

In addition to a direct contribution to the thinning of the ozone layer, which causes an increase in the solar heat flux reaching Earth, this destruction of stratospheric ozone induces by a pendulum effect an increase in tropospheric ozone in the lower atmosphere. Ozone, like CO_2, is a well-known greenhouse gas, that is, it captures the thermal radiation reflected by Earth, which contributes to the warming of the atmosphere and, eventually, to the warming of Earth's surface.

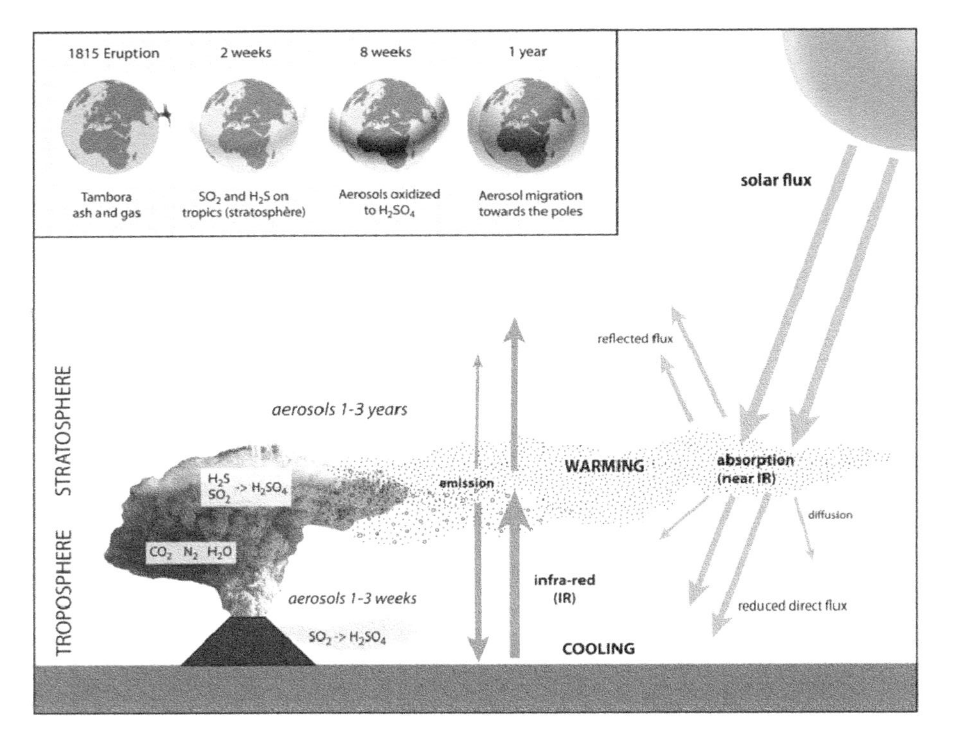

Figure 2.2. *Schematic representation of the influence of volcanic sulfate aerosols on the terrestrial radiation budget (adapted from Robock (2000)). Inset: dispersion of volcanic gases and aerosols from the Tambora eruption in 1815–1816 (Brönnimann and Krämer 2016). Adapted from Brönnimann and Kramer (2016). For a color version of this figure, see www.iste.co.uk/lenat/hazards.zip*

An example of such a volcanic climate warming process is found at the Cretaceous-Tertiary boundary, 66 million years ago, during the formation of the Deccan Trapps, a large basaltic province of 500,000 km² and 2,400 m thick, covering the entire western Indian subcontinent and associated with the Mascarene hotspot. The amount of CO_2 released into the atmosphere during the formation of this volcanic province is estimated at 36,000 megatons, an increase in the partial pressure of atmospheric CO_2 of about 1,000 ppmv according to the models considered[1]. Until all of this volcanic CO_2 was consumed during basalt weathering processes during the 1–2 million years after the formation of the Deccan Trapps, it resulted in an exceptional rise in global surface temperatures of between 2°C and 4.5°C (Dessert et al. 2001) (see Figure 2.3). To say that the rise in temperature following this particularly intense volcanic episode is responsible for the mass extinctions marking the Cretaceous-Tertiary boundary would probably be too hasty. However, it is clear that in an Earth system already strongly disturbed by the major impact of the Chicxulub asteroid, such a modification of the conditions of life on Earth had a lasting impact on biodiversity and contributed to the decline of certain species such as the large saurians.

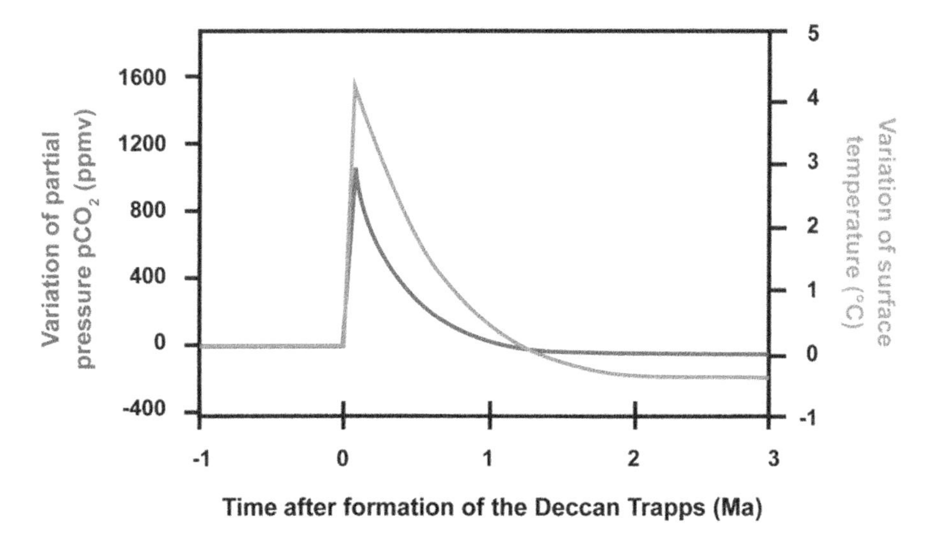

Figure 2.3. *Climate impact model related to the emplacement of the Deccan Trapps. Modified from Dessert et al. (2001). For a color version of this figure, see www.iste.co.uk/lenat/hazards.zip*

1 Such an increase is colossal: it corresponds to more than twice the current CO_2 content of Earth's atmosphere, measured at 416 ppmv in 2020.

COMMENT ON FIGURE 2.3.– *From the beginning T_0 of the eruptive sequence, the CO_2 injected in the atmosphere generates a significant increase of the partial pressure pCO_2 (red curve). At the end of the eruption, pCO_2 gradually decreases to its pre-eruption level. At the same time, this increase in pCO_2 leads to an increase in global surface temperatures by radiative forcing (blue curve). As soon as the CO_2 starts to be consumed during the process of alteration of the rocks put in place, the temperature falls in turn and even reaches values lower than those initially present before T_0.*

2.2.2.3. *Local gas hazards: an invisible but real threat*

Catastrophic eruptions of this kind are rare, so major changes induced by volcanic gases on a global scale are also rare.

On a regional or even local scale, volcanic outgassing remains an important, though often overlooked, hazard. Moreover, since gases are usually odorless and colorless, it is a sneaky hazard for which populations are generally ill-prepared. However, most of the volatile species that make up volcanic gases are known to be toxic to humans and other living beings, which make them dangerous.

The second largest constituent of volcanic gases, carbon dioxide (CO_2) can be lethal through a hypoxia effect starting at a volume concentration of 10% (increasing CO_2 concentration in inhaled air induces a decrease in oxygen content) but it itself becomes lethal when its concentration exceeds ~35% in air (Mastrodicasa et al. 2018). In practice, such levels are only very rarely reached in volcanic gases, which are predominantly dominated by water vapor, even in hot-spot settings (see Table 2.1). On the other hand, CO_2 is a denser gas than other atmospheric components and thus has a natural tendency to accumulate in topographic depressions and other cavities, as evidenced, for example, by the famous Chien Cave in Chamalières (Puy-de-Dôme, France) and the eponymous Pozzuoli Cave in Italy. In these two rather similar caves, CO_2 of magmatic origin accumulates by density contrast at the topographic low point of the cave, so that a dog breathing a few decimeters above the ground quickly asphyxiates while its master does not even feel respiratory discomfort. If these two caves were mainly tourist attractions, the same process of CO_2 accumulation could constitute a particularly formidable hazard in other cases. The disasters of the Dieng Plateau (Indonesia, 1979, hydrothermal CO_2 flow released during a phreatic eruption, 142 deaths), Lake Monoun (Cameroon, 1984, intracrateric landslide causing the sudden release of a deep CO_2-enriched layer of the lake, 37 deaths) and Lake Nyos (Cameroon, 1986, limnic eruption with sudden release of a CO_2-rich meromictic layer, 1,746 deaths) are dramatic examples of the volcanic hazard associated with the gas phase. They remind us that deep crater lakes can pose a serious threat despite their apparent tranquility; many lakes have been monitored for CO_2 content in their deep waters (Lake Pavin, France; Laacher See, Germany; Mount Gambier, Australia). If the deep meromictic layers reach CO_2 saturation, a Nyos-type episode is then possible, the means of prevention being to mechanically purge the CO_2 by a drainage system from the bottom of the lake causing an artificial geyser at the surface, as in the case of the "pipe of Nyos" (see Figure 2.4).

CO_2 is obviously not the only potentially lethal compound in volcanic gases. Carbon monoxide (CO), often associated with CO_2, is suspected to be the cause of death of Werner Giggenbach (1937–1997, internationally renowned volcanologist and gas geochemist) and five other people on the Tavurvur crater in the Rabaul caldera (Papua New Guinea). Sulfur gases, SO_2 and H_2S, have been implicated in the deaths of several dozen people on European (Etna, Santorini), Indonesian (Papandayan), Japanese (Aso, Kusatsu) or Central American (Santa Maria) volcanoes according to the catalog established by Siebert et al. (2010).

Figure 2.4. *Lake Nyos pipe (credit M. Halbwachs)*

COMMENT ON FIGURE 2.4.– *The deep waters are initially pumped mechanically and rise in a draining system connected to the surface. The pressure drop generated by the ascent of these deep waters induces an exsolution of CO_2 in the tube. The less dense gaseous water then rises naturally to the surface (i.e. without pumping) and carries in its wake the deep water that has not yet been outgassed, forming a continuous geyser at the surface of the lake.*

Even without directly leading to the death of exposed people, acidic volcanic gases (mainly sulfur species and halogenated species such as HCl and HF) have a detrimental effect on living things and the environment in general. The recent eruption of Kilauea in Hawaii in 2018 generated a vog (a British neologism from volcanic fog) at the Pahala settlement with SO_2 concentrations above 500 ppbv in several instances (Tang et al. 2020). Similarly, during the 2014 eruption in Holuhraun (Iceland), the Icelandic civil protection indicated, for the morning of October 30, SO_2 concentrations up to 4 mg/m^3 in Akureyri, the second largest city in the country[2]. An in-depth epidemiological study on the effects of acidic volcanic gases has yet to be carried out, but studies in the workplace[3] show a strong impact on the skin (redness and dermatitis that can lead to skin lesions), on the internal mucous membranes (oral, esophageal and stomach burns) and on the ocular system (corneal burns that can lead to loss of sight). The main route of entry of these compounds into the body, however, is inhalation. The main effects affect the respiratory system with, among other things, acute or chronic diseases, coughing, dyspnea, pharyngitis, bronchitis or asthma. From an environmental point of view, the impact of these acid gases on the vegetation is also remarkable with an acidification of the soils and a fumigation of the foliage, which can lead to plant necrosis. If geographical areas downwind are chronically exposed to volcanic gas, vegetation can be totally destroyed or at least experience a significant decrease in productivity (Winner and Mooney 1980). In the case of food crops, the social impact, in addition to the health impact, is obviously immediate for the local population (Delmelle et al. 2002).

2 The atmospheric concentration of SO_2 is governed by European standards with a regulatory threshold set at 125 μg/m^3 (i.e. ~50 ppbv) as a daily average (not to be exceeded more than 3 days per year) or at 350 μg/m^3 (~135 ppbv) as an hourly average (not to be exceeded more than 24 hours per year).

3 See, for example, www.inrs.fr.

2.2.2.4. *Poisoning and pollution by volcanic gases: the local biosphere at risk*

Soil and plant poisoning, especially with fluorine, can add to their acidification and amplify the effects of the gas hazard, as illustrated by the 1783 Laki eruption (Iceland). Approximately 8 megatons of hydrogen fluoride were dispersed into the environment during this eruption, mostly adsorbed on the surface of the ash. Ingestion of this fluoride-enriched ash by grazing animals caused bone and dental fluorosis, leading to the death of more than 60% of Iceland's livestock and, indirectly, to the death of 20% of the Icelandic population due to starvation and malnutrition (Thordarson and Self 2003).

Finally, the atmospheric injection by volcanoes of many other toxic chemical elements (heavy metals – Cd, Hg, Tl, Pb among others – and metalloids, e.g. As and Se) or radioactive elements (Rn and Po) is now well documented and shows that outgassing is an important, even preponderant, source in the cycle of some of these elements (Allard et al. 2016; Gauthier et al. 2016). The atmospheric dispersion of these pollutants, their fate and environmental residence time, their ability to be assimilated by living organisms and the damage they may cause to these organisms remain open questions to date. Let us hope that future systematic and epidemiological studies in relation to these toxic elements will complete our understanding of the hazard related to volcanic gases and outline the most appropriate responses for the protection of populations living in volcanic areas.

2.2.3. *Lava flows*

2.2.3.1. *Characteristics of lava flows*

Lava flows are the product of an outpouring of molten volcanic rock that forms during effusive eruptions. Lava flows are gravity currents propagating under their own weight, slowed down by the topography or by their own viscosity (which increases with cooling). Their setting (velocity, rheological behavior, morphology and distance) depends mainly on the effusion rate (volumetric flow at the blowhole), the volume emitted, the topography and the thermorheological properties of the lava. The higher the effusion rate and volume, the longer the flow. The lower the viscosity and steeper the slope, the faster the flow will spread. A lava flow stops spreading when the supply of lava stops or when the cooling is such that its viscosity reaches a limit that prevents it from advancing.

Different types of lava flows can be distinguished according to their composition, morphology and dynamics of setting up. Lava flows poor in silica (e.g. basaltic flows) are the hottest and most fluid and can reach speeds of several km/h. These flows are found on effusive volcanoes, such as Hawaii (USA), Piton de la Fournaise (Reunion Island), Nyiragongo (DRC) or Etna (Italy). The most fluid basaltic flows, of the pāhoehoe type, flow as lava lobes when the flow rate is low (see Figure 2.5(a)). These lobes are a few tens of centimeters to a few meters thick and have a smooth surface due to the extension of a thin, still ductile crust on the surface. When the lava is cooler, crystalline, and viscous, it flows as ‘a’ā-type with a thicker, fragmented and scoriated cold layer on the surface (see Figure 2.5(b)). At higher flow rates, the flows become channeled, forming true torrents that can extend over several tens of kilometers (see Figure 2.5(c)). The formation of tunnels isolates the lava and allows it not to cool and thus to reach great distances. The more siliceous compositions (andesite, dacite and rhyolite) form so-called block flows, which are colder and more viscous; their speed does not exceed a few meters per day. Their thickness is generally several meters to several tens of meters and their surface is made of angular blocks that can reach sizes of the order of 1 m (see Figure 2.5(d)). These flows are often observed on the flanks of stratovolcanoes (from subduction volcanism). When the viscosity is very high, these lavas cannot flow and accumulate to form a dome. Lava domes are vertical extrusions that can reach several tens of meters in height above the central crater of a volcano. Extrusion can occur rapidly in a single surge or in a succession of events. In some cases, these domes may begin to flow to form block flows or destabilize and collapse, forming block and ash avalanches.

2.2.3.2. *Hazards associated with lava flows*

Lava flows are one of the most common volcanic hazards worldwide; however, they rarely pose a risk to people (Harris 2015). One of the deadliest examples is the 1977 and 2002 Nyiragongo eruptions in the Democratic Republic of Congo (DRC). These emitted extremely fluid and rapid lava flows that reached the city of Goma, located 10 km to the south, within hours. During these two crises, several dozen or even hundreds of people died from carbon dioxide asphyxiation, in the collapse of buildings covered by flows or shaken by the earthquakes, or in the explosion of fuel tanks invaded by a flow. In the 2002 crisis, approximately 120,000 people (15% of the city) were left homeless, and one-third of the airport runway was covered by lava (Tedesco et al. 2007).

Figure 2.5. *Different types of lava flow morphology. For a color version of this figure, see www.iste.co.uk/lenat/hazards.zip*

COMMENT ON FIGURE 2.5.– *a) Pāhoehoe lobes, Pu'u 'Ō'ō, Kilauea, Hawaii, USA (temperature measurements (credit M. Patrick 2016)). b) Active 'a'ā lava front (2 m thick) advancing over an older 'a'ā flow, Piton de la Fournaise, Reunion Island, France (credit O. Chevrel 2019). c) Channelized lava flow, Kilauea, Hawaii, USA (credit USGS 2018). d) Block lava flow, Arenal, Costa Rica (credits William Melson 1968).*

Figure 2.6. *Houses buried in a lava flow during the 2014 Fogo (Cape Verde) eruption (photo: R. Paris)*

Casualties from lava flows are minimal compared to other volcanic events (they account for only 0.2% of total volcanic hazard-related deaths (Harris 2015)), but the damage they cause can be significant. The covering of cropland and infrastructure has lasting impacts on local economies. For example, in June 2018, the effusive eruption of Kilauea in Hawaii (Neal et al. 2018) resulted in no fatalities but destroyed more than 700 homes and caused estimated losses of more than $800 million.

On the flanks of volcanic islands, it is common for lava to reach the coastline. When the lava comes into contact with sea water, the instantaneous thermal expansion of the water causes violent explosions with the projection of ashes and blocks and the formation of a plume. This plume, called "laze" (lava and haze, fog), is loaded with harmful gases and aerosols (HCl, CO_2, NO_2 and SO_2) and can induce acid rain, which represents a significant danger for the surrounding populations and farms. The accumulation of lava on the submerged slopes of the island forms unstable terraces that can collapse at any time and form local tsunamis.

Lavic effusion on a substrate saturated with water, snow or ice can also cause phreatomagmatic explosions along a lava flow and form small pyroclastic flows. In some cases, a lava flow can cause partial melting of the snow or ice substrate, generating mudflows (lahars) that can quickly reach great distances and raise river levels.

Block lava flows can also be a significant hazard because their fronts are unstable, causing falling blocks and surges of ash and blocks. Lava domes are particularly dangerous because of their instability, as are the fronts of block flows. Gravity collapse of a dome can generate debris avalanches and, through rapid decompression of the underlying magma column, explosive activity often associated with pyroclastic flows and a plume (see sections 2.2.4 and 2.2.6). The accumulation of pyroclastic deposits is then conducive to the formation of lahars (see section 2.3.1).

Finally, some effusive eruptions with exceptional volumes of lava result in large discharges of gas into the atmosphere that can have an impact on the environment and climate. For example, the Laki eruption in Iceland in 1783 formed a lava field of 565 km^2 (that is, 15 km^3) of lava in only 8 months. The amount of SO_2 released into the atmosphere caused climatic disruption for a few years affecting all of Europe and causing famines (see section 2.2.2.1). On a much larger scale, the large basaltic provinces from the hotspots were formed by the emission of lava flows that could cover more than 10,000 km^2 for volumes of several thousand km^3 over relatively short periods. Such events, due to associated outgassing and aerosol production, have had a global climate impact and have been associated with mass extinctions (White and Saunders 2005) (see section 2.2.2.1).

2.2.3.3. *Lava flow monitoring and control*

During an eruption, lava effusion can be followed by various means: directly on the ground, by overflights (helicopter, ULM, UAV) and by remote sensing on the ground, from the air or from space. In the field, volcanologists obtain the exact position of the vent and the lava front, measure the flow velocity, take samples and measure the temperature. The sampling of the molten lava can only be done if the flow can be approached. This is only possible if it is pāhoehoe (see Figure 2.5(a)) or if it is well channeled between stable levees or through an opening in the roof of a lava tunnel. Approaching a moving boulder flow is impossible because of falling boulders. The surface temperature can be measured with a thermal camera in the field or during a flyover. The overflight gives an overall view of the eruption and, thanks to photogrammetry, allows the topography of the lava field to be reconstructed (Derrien 2019) and thus the volume of lava emitted to be estimated. Remote sensing techniques are essential for monitoring eruptions when they cannot be approached. There is a wide variety of possible measurements. Those most commonly used during effusive eruptions are images acquired in the visible and infrared spectrum in order to obtain an overview of the extent of the flows, and also to measure the thermal radiance and thus to deduce the lava flux and the emitted volume (Gouhier et al. 2016; Harris et al. 2016; Coppola et al. 2017) (see also Chapters 3 and 4 of Volume 2). Recently, the tracking of a flow can also be done by radar interferometry, which provides a precise contour of the extent of the deposits (Dietterich et al. 2012; Harris et al. 2019; Hrysiewicz 2019) (see Figure 2.7; and Volume 2, Chapter 2). Today, there are a multitude of sensors on different satellites, and by making the best use of their overflight of the volcano (depending on the orbits), one can practically track an eruption at close intervals (see Volume 2, Chapter 3).

Meanwhile, the internal properties of the lava (petrology, chemistry and density) cannot be determined directly in the field but require laboratory analysis. In rare cases, viscosity can be measured in the field using appropriate instrumentation such as a field viscometer (Chevrel et al. 2019).

With the exception of the Goma example cited above, in most cases the advance of the flows is slow enough to allow people to evacuate. However, non-mobile structures (buildings, roads, etc.) cannot avoid burning and burial and agricultural land is rendered barren. For evacuation of populations to be possible, evacuation and relocation plans must be developed in advance at the local and regional (or even national) levels and be well known to the population at risk. In rare cases, people have tried to intervene to control the path of lava flows. *Dams* to hold back or slow lava flows have been constructed several times in the past, such as on Etna in Italy and in Hawaii, Japan and Iceland (Harris 2015). *Watering* the lava front in order to cool it and thus increase its viscosity to freeze it has also been attempted. However, the advance of a flow can only be slowed for a limited time. If the source continues to be active and the flow of lava consequent, then the barrier will be exceeded. Moreover, if the barrier breaks, this can have even more serious consequences because the lava accumulated upstream will pour out even faster than the original flow. The intensive drenching of a flow on Heimaey Island in Iceland in 1973 was a success. The operation, which consisted of pouring 100 m^3 of water per second over a 500 m long front, allowed the lava front to solidify and thus prevented the blockage of the island's harbor.

The *detour* of the flows through the construction of detour dykes proved to be very useful on Etna in 2001. A succession of dikes was erected to redirect the flow and spared busy tourist areas (Barberi et al. 2003).

Bombing and explosives have also been used to break the walls or roofs of lava tunnels, to pierce channel levees or to partially destroy the cone formed at the vent so as to make the lava flow in a different direction. This technique was used in the 1930s–1940s in Hawaii and in the 1980s–1990s on Etna.

It is possible to determine the most likely *trajectory* that lava might take during an eruption using numerical models (see Figure 2.8). Since a lava flow is a gravity flow, its trajectory follows the steepest slope. Simple probabilistic models can thus define the entire area likely to be covered by the lava. However, these models do not take into account the thermal and rheological properties of the lava and therefore do not predict how far the lava may extend (Wadge et al. 1994; Favalli et al. 2005).

By reconstructing the past activity of a volcano, it is possible to establish the statistical relationship between the flow rate and the final length of the flows. In addition, when temperature measurements and sampling are made along a flow, it is possible to estimate the evolution of thermo-rheological properties with cooling, and thus to infer the distance at which the lava is expected to freeze (Harris and Rowland 2001). By associating these models with the potential trajectory, it is possible to draw up lava flow impact hazard maps (Rowland et al. 2005). During an effusive crisis, a map of the probability of inundation and the maximum distance the flow could reach can then be quickly developed and communicated to authorities (Harris et al. 2019) (see Figures 2.9 and 2.10).

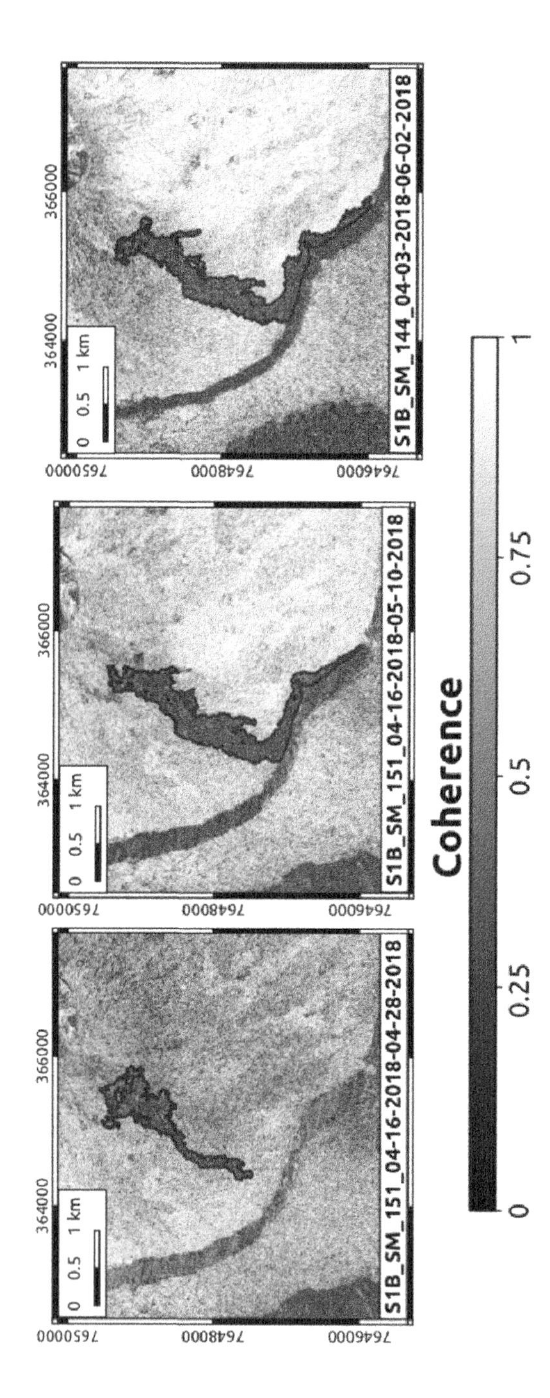

Figure 2.7. *Tracking a flow by radar interferometry during the April 27, 2018, eruption at Piton de la Fournaise (Harris et al. 2019). For a color version of this figure, see www.iste.co.uk/lenat/hazards.zip*

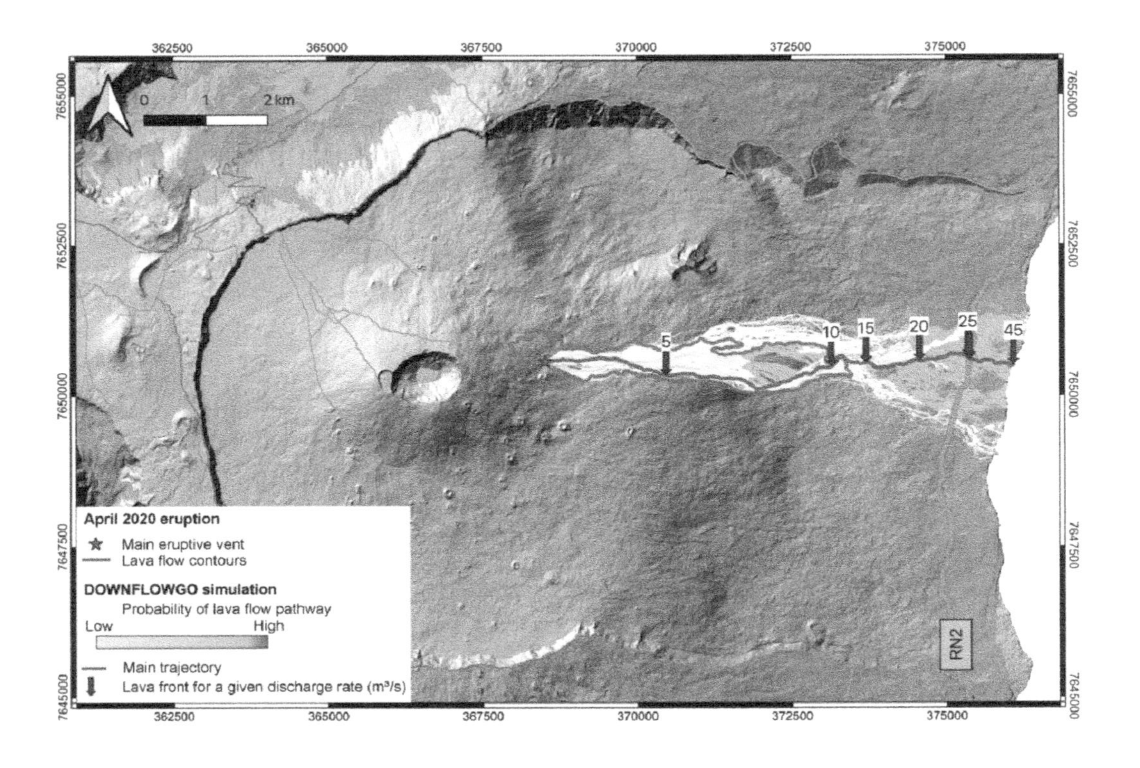

Figure 2.8. *Numerical simulation of lava flow pathway probability obtained for the April 2–6, 2020, eruption at Piton de la Fournaise (from Bulletin of the Observatoire Volcanologique du Piton de la Fournaise, Bulletin ISSN 2610-5101) with the DOWNFLOWGO model (Harris et al. 2019). For a color version of this figure, see www.iste.co.uk/lenat/hazards.zip*

COMMENT ON FIGURE 2.8.– *The main (most probable) trajectory is shown in red and the blue arrows indicate the position at which the lava should freeze for a given flow rate (associated number in* m^3/s*). The pink line represents the final contour of the lava flow.*

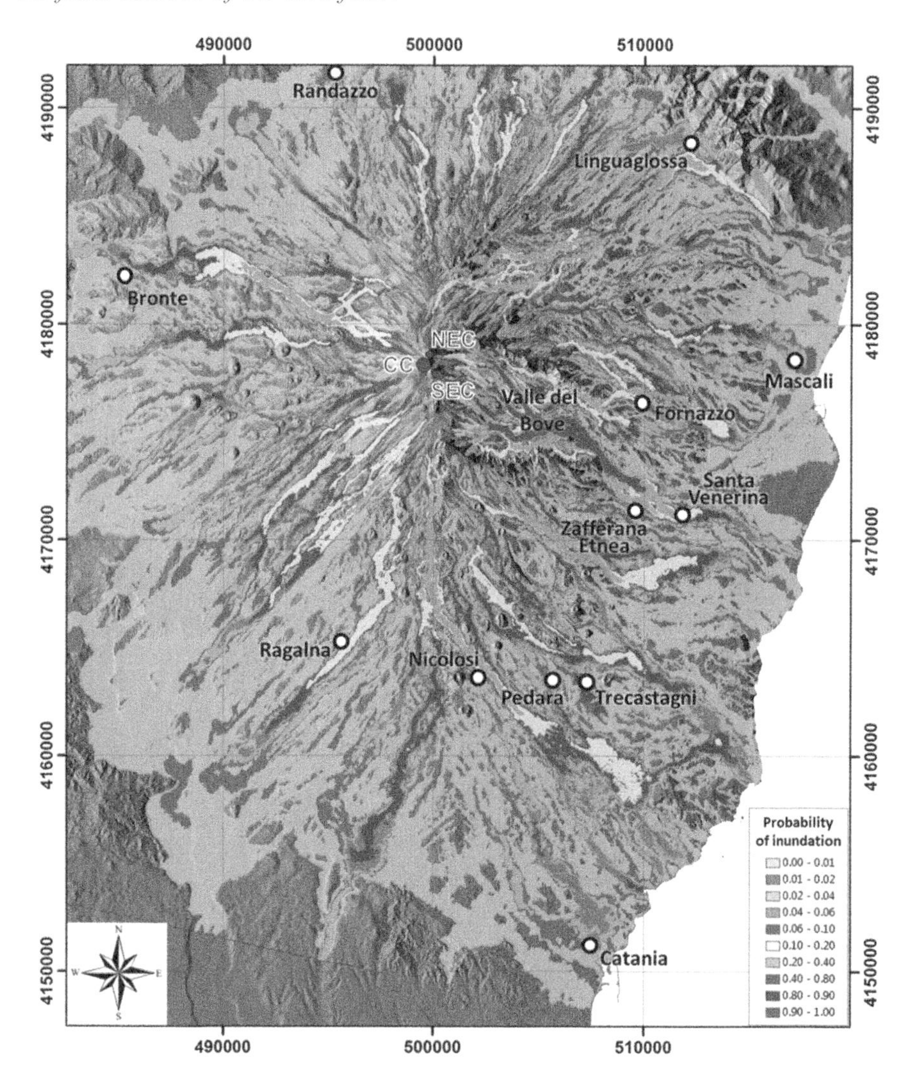

Figure 2.9. *Lava flow hazard maps (probability of lava flooding), Etna, Italy (Del Negro et al. 2013). For a color version of this figure, see www.iste.co.uk/lenat/hazards.zip*

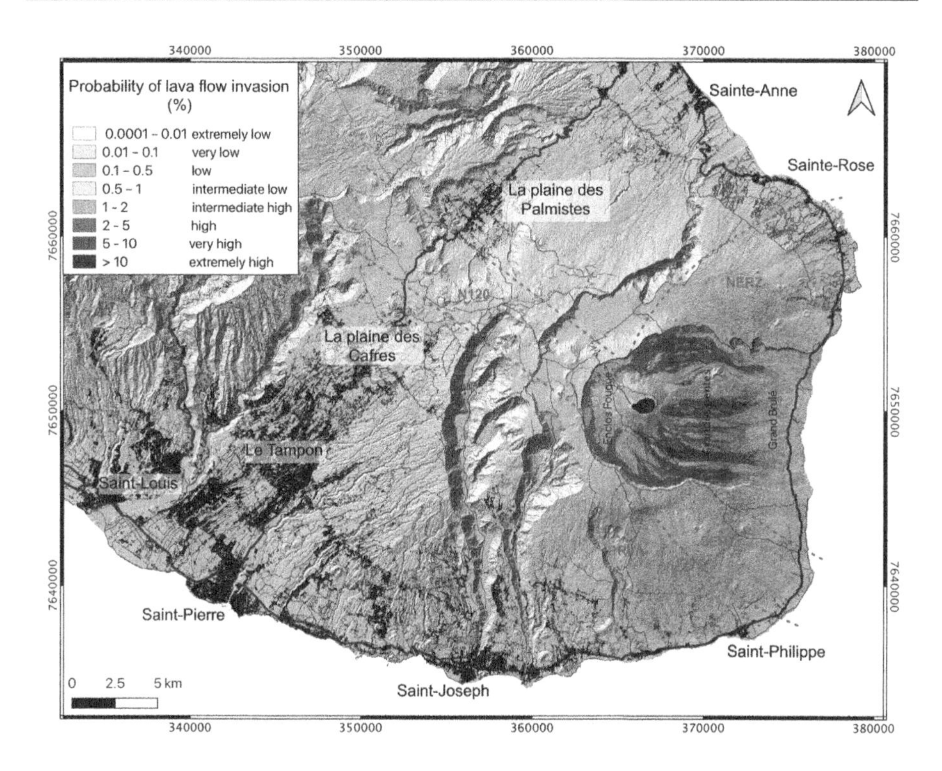

Figure 2.10. *Lava flow hazard maps (probability of lava flooding): Piton de la Fournaise (Reunion Island) (Chevrel et al. 2021). For a color version of this figure, see www.iste.co.uk/lenat/hazards.zip*

Other models, called deterministic, are based on mass conservation and consider the behavior of the flow as a whole using a maximum value (viscosity or yield point) above which flow is impossible (Dragoni et al. 1986; Miyamoto and Papp 2004; Del Negro et al. 2005; Bernabeu et al. 2016; Kelfoun and Vallejo Vargas 2016). The numerical resolution of these models is more complex and often requires longer computation time, which may not make them suitable during crisis management.

On most effusive volcanoes, lava flows are not necessarily emitted from the summit but from fissure openings on the flanks of the edifice. One of the main issues is therefore to know the distribution of the probability of opening a crack and the extent of lava flows associated with these openings. When this is known, it is then possible to establish lava flood hazard maps (Del Negro et al. 2005; Mossoux et al. 2019) (see Figure 2.9).

Another issue is the a priori knowledge of the volume of lava and the flow rate emitted during an eruption (see Volume 2, Chapter 3) and the thermorheological evolution of the lava during its effusion. This knowledge will allow anticipating with more precision the dynamics of a flow, its speed, its trajectory of effusion, its duration and its final dimensions.

2.2.4. *Tephra*

Volcanic explosions produce and eject tephra of various sizes and gases. Tephra are classified into three main classes: bombs (>64 mm), lapilli (between 2 and 64 mm) and ash (<2 mm). Tephra can be juvenile (belonging to the magma of the eruption) or accidental (torn from the rocks of the conduit and the edifice – also called "lithic"). Depending on the type and intensity of the eruptions, the products will have different paths through the atmosphere, from ballistic fallout around the vent to plumes rising to high altitudes (see Figure 2.11).

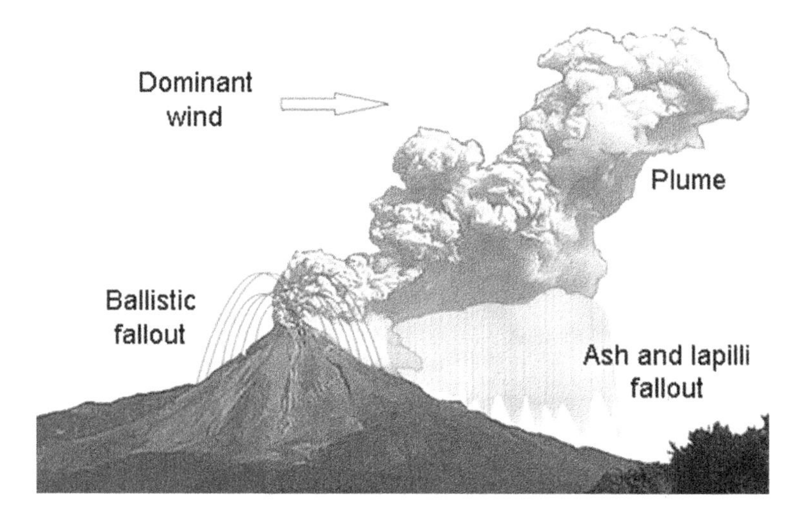

Figure 2.11. *Main types of tephra*

2.2.4.1. *Characteristics of volcanic plumes*

Volcanic plumes consist of a mixture of silicate particles and gases that rise vertically or sub-vertically into the atmosphere and then disperse horizontally under the influence of winds (see Figure 2.12). The tephra will fall back in various processes, depending on their size, density and plume momentum, to form deposits (Carey and Bursik 2015). An iconic plume is that produced by the May 18, 1980 eruption of Mount St. Helens, which reached an altitude of 30 km, traveled thousands of kilometers across the American West (Holasek and Self 1995), and covered tens of thousands of square kilometers with tephra (Sarna-Wojcicki et al. 1981). Also noteworthy is the plume from the Icelandic volcano Eyjafjallajökull, which in April 2010, due to sustained southeasterly winds, paralyzed northern European airspace (Bonadonna et al. 2011). Despite its persistent nature (39 days), the plume was modest in size, varying between 3 and 10 km in altitude (Gudmundsson et al. 2012).

Figure 2.12. *Photograph of an ash plume showing the vertical column and its horizontal spreading (e.g. Redoubt volcano, 1990, Alaska, photo: USGS)*

A plume is initiated by the violent and rapid decompression of gases in the eruptive vent, which produces a jet of material and gases. The heating of air entrained in a pyroclastic flow can also generate a plume (see Figure 2.13). Its ascent into the atmosphere is controlled by its buoyancy, that is, the vertical thrust force produced by the high-temperature gas-particle mixture, whose density is lower than that of the ambient air. This ascent phase generates a convective column in which the ambient air is drawn in and heated and then expands, thus increasing the diameter and therefore the volume of the column. If the amount of air assimilated is small relative to the mass of ash, the plume can collapse under its own weight and generate pyroclastic flows (Wilson et al. 1980; Dufek et al. 2015). Vertical plumes, known as strong plumes, are observed when the column rise velocity is greater than the horizontal wind velocity, and leaning plumes, known as weak plumes, are observed when the column rise velocity is significantly lower than the wind velocity (see Figure 2.13). The plume rises through the atmosphere until its density is equivalent to that of the ambient air (neutral buoyancy level). Nevertheless, the maximum height of a plume (Ht) may exceed the neutral buoyancy level (Hb), due to the inertial forces of convective mixing, that is, when the plume velocity is non-zero at Hb. When it reaches the neutral buoyancy level, the plume spreads horizontally to form an umbrella within which gravity forces dominate.

The plumes generated at the vent are initially controlled by various parameters such as the magma ascent rate, the proportion of gas, the nature of the liquid, the depth and efficiency of fragmentation and by the ejection rate of the gas-particle mixture into the jet, inherited from the momentum produced by the depressurization of volcanic gases. The initial jet transforms into a convective column due to the entrainment of ambient air, which decreases the density of the gas-particle mixture. Co-flow plumes are formed with the decrease in flow density induced by sedimentation of some of the transported material and entrainment of ambient air into the flow body (Engwell and Eychenne 2016). A mixture of gas and fine particles then rises in a convective column.

The granulometry and density of the tephra, as well as the dynamics and size of the plume, control their fallout. Bombs and blocks settle near the source following ballistic trajectories. The lapillis are, depending on their size, partly entrained in the plume. The size of the deposited lapillis decreases with distance. Ash, and in particular fine ash with a diameter of less than 100 μm, is carried along in the plume and can travel thousands of kilometers. However, larger than expected ash fallout can be observed at relatively small distances from the source. These anomalies can be explained by aggregation and accretion of ash that forms larger tephra (Brown et al. 2012) and by local gravitational instabilities in the plume (Scollo et al. 2017) that amplify sedimentation (Gouhier et al. 2019).

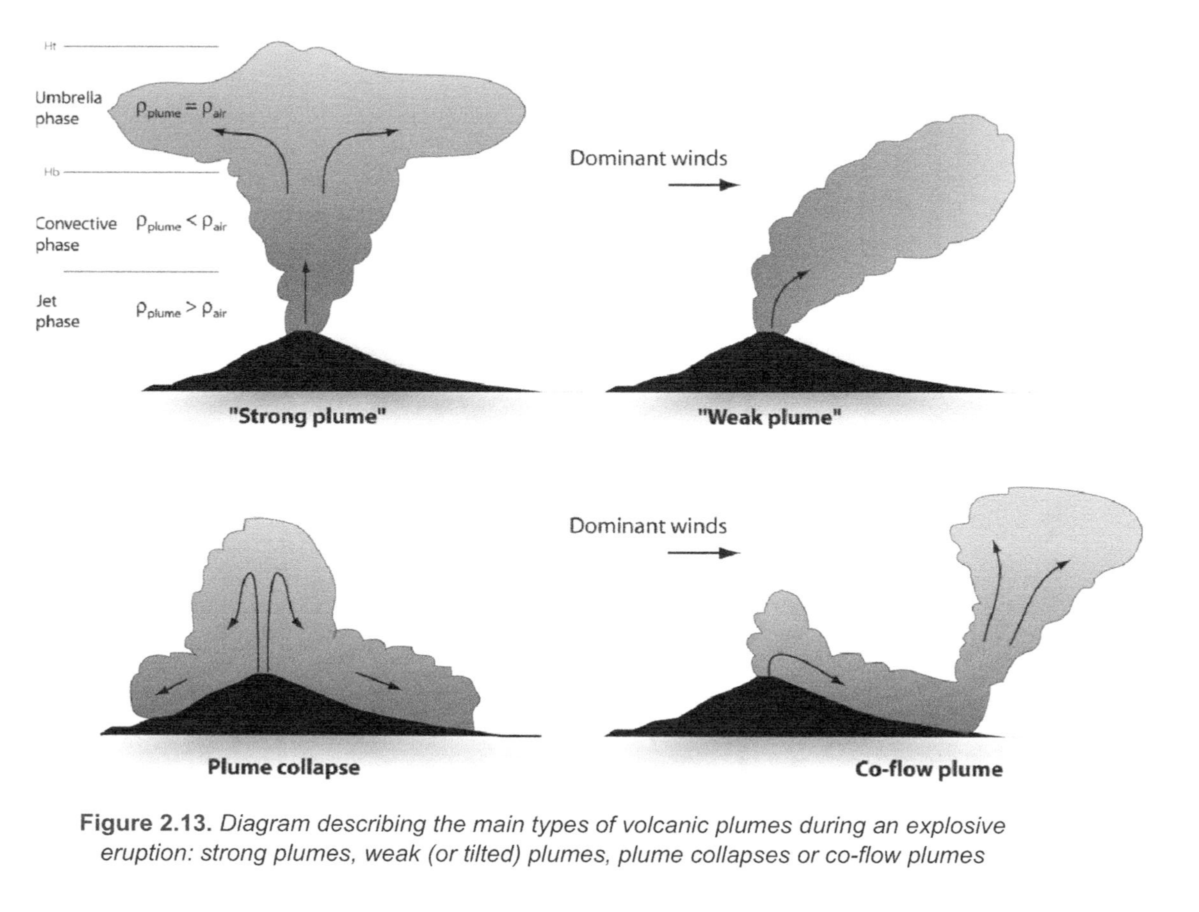

Figure 2.13. *Diagram describing the main types of volcanic plumes during an explosive eruption: strong plumes, weak (or tilted) plumes, plume collapses or co-flow plumes*

Fallout deposits cover large areas. Their thickness decreases with distance from the source. They can be used to reconstruct the dynamics of past volcanic eruptions and are also good markers (tephro-chronology) for paleoenvironmental studies.

2.2.4.2. *Hazards associated with tephra plumes and fallout*

The hazards associated with bombs and blocks are restricted to impacts in the vicinity of the vent, up to a few kilometers for the most explosive activities. Fires can also be caused by hot ejecta. The area affected by ash and lapilli fallout depends on several parameters: the style and magnitude of the eruption, wind conditions at different altitudes and the characteristics of the plume itself. There is an evolution of ash grain size and deposit thickness with distance from the vent (see Volume 3, Chapter 2).

Figure 2.14. *Ash deposits from the eruption of the Tungurahua volcano (Ecuador) in 2006 (photographs: Raphaël Paris)*

In terms of impact, the main effects on infrastructure and networks (Wilson et al. 2012, 2015) are:

– *Roof collapses* (see Figure 2.14) when the load exceeds the strength of a roof or frame. The density of dry ash ranges from about 400 to 1,000 kg/m^3 but can be increased by 50%–100% if it is wet.

– Disturbance of *electrical installations and appliances*. As wet ash conducts electricity, it can cause short circuits, both on power lines and transformers and on any installation in contact with the ash.

– *Other electrical or electromagnetic effects* exist, such as lightning (a frequent phenomenon in plumes) or the diffusion and absorption of radio signals.

– Damage to *engines of machinery or vehicles*, by clogging of their filters or abrasion of moving parts.

– *Traffic* disruptions. Ash accumulation can make roads impassable, and visibility can become critically reduced with airborne ash.

– *Pollution* of open water supplies (or water supplied by open systems), which may become acidic and turbid and therefore unfit for consumption.

Tephra plumes have a significant impact on air traffic. Between 1953 and 2009, there were 94 confirmed cases where aircraft encountered volcanic plumes and suffered damage (Guffanti et al. 2011). In nine cases, there was temporary engine shutdown. Although there were never crashes, the aircraft often suffered significant damage.

When an aircraft passes through a volcanic plume, the main effects are:

– abrasion of windows, fuselage, wings and headlights;

– obstruction of probes and electromagnetic and electrical disturbances;

– pollution of the air conditioning system;

– damage to the engines.

The combustion zone of modern engine turbines is around 1,200°C–1,500°C, while the glass particles in the plumes melt between 700°C and 1,000°C. Drops of silicate liquid are formed and projected onto the turbine blades, where they are frozen (soaked) and form a hard layer that will seize the turbines, eventually causing the engines to shut down. These incidents have led to the development of specific tools to ensure the safety of commercial and other flights in the event of a volcanic eruption.

Effects on vegetation are variable depending on vegetation type, climate, ash chemistry and ash thickness (Wilson and Kaye 2007). A relatively thin layer of ash (<20 mm) may prevent animal grazing and restrict photosynthesis. From a thickness of about 30 cm, much of the vegetation disappears.

On a global scale, volcanic plumes have a climatic impact when they reach the stratosphere or when intense tropospheric plumes occur over long periods of time (Robock 2000). Their effects on climate are due to a combination of gases, aerosols and ash.

The classic health effects are irritation of the nose and throat, due to the combined effect of mechanical irritation and the acidity of the ash. But in some eruptions, the ash particles can be so fine that they can penetrate deep into the lungs. Their particle size determines their penetration into the lungs. Particles smaller than 100 μm cause upper respiratory tract irritation, whereas finer particles (<10 μm) penetrate the lungs and can cause asthma or bronchitis. The finest particles (<4 μm) can be deposited in the alveoli and potentially cause irreversible chronic respiratory pathology (Horwell and Baxter 2006). Eye and skin irritation and fluorosis poisoning symptoms are also noted (D'Alessandro 2006). Fluoride is essential for growth in humans and many animals but becomes toxic in high doses. Ash can carry large quantities of fluorine, depending on the nature of the magma and the style of the eruption. This will end up on pastures, fruits and vegetables and in water. Human and livestock infections because of fluorine are regularly linked to eruptions; for example, Mali (Iceland) in 1783, Hekla (Iceland) in 1970 and Lonquimay (Chile) in 1988–1990. Finally, the volcanic glass filaments (Pele's hair) formed in lava fountains are carried by the wind; in addition to the micro-injuries of the skin and eyes that they can inflict, they constitute a danger for the digestive system of animals that ingest them in the pastures.

2.2.4.3. *Prevention of tephra-related hazards*

Protection against falling bombs and blocks is fairly simple, since it is sufficient to evacuate the area near the volcano. Concrete shelters can also be provided. This is the case at the top of Stromboli and around Sakurajima where the population is regularly exposed to tephra falls (school children wear helmets to go to school). For ash fallout, the problem is more complex. This has led the scientific community to publish booklets for the public and the authorities to manage this hazard. The International Volcanic Health Hazard Network[4] (IVHHN) website provides information and guidelines for minimizing the risks associated with volcanic ash and gas, including booklets such as "How to Prepare for and Deal with Falling Ash" and "How to Protect Yourself from Breathing Volcanic Ash".

Another prevention approach is plume monitoring and plume trajectory and fallout modeling. Remote plume monitoring is discussed in the chapters dedicated to ground- (see Volume 2, Chapter 4) and space-based remote sensing monitoring (see Volume 2, Chapter 3). Monitoring and analysis (extension, thickness, composition and granulometry) of deposits are also crucial for hazard zonation both in times of crisis and for characterizing past activities (see Volume 3, Chapter 2).

Models of ash plume and fallout dispersion have been developed for decades. Recent reviews of these approaches can be found by Bonadonna and Costa (2013) and Jenkins et al. (2015). These models tend to provide an estimate of fallout thickness and spatial concentration of ash in the atmosphere. This is a developing field because the physical processes involved are complex. One of the difficulties in applying these models is that the source conditions (flux, velocity and composition) are difficult to determine. Some remote sensing techniques, such as the use of near-vent Doppler radars, can help to define the initial conditions (see Volume 2, Chapter 4).

Plume dispersion models are used for aviation safety. Following the Pinatubo eruption in 1991, the International Civil Aviation Organization created a network of nine VAACs (Volcanic Ash Advisory Centers) covering almost the entire globe. Their function is to provide real-time alerts for aviation. Their action is based on:

– information provided by volcanological observatories and other institutions;

– meteorological data and observations from satellites;

– information from aircraft pilots;

– plume dispersion models.

4 https://www.ivhhn.org/.

Until now, it has been difficult to visualize or detect plumes on board aircraft. Specific equipment is being developed. The AVOID (Airborne Volcanic Object Identifier and Detector) system includes two infrared cameras installed either on the wingtips or on the fuselage. This system is capable of detecting ash clouds up to 100 km ahead. At normal flight speed, this gives the pilots 7–10 minutes to divert and avoid a plume.

2.2.5. *Atmospheric pressure waves*

The so-called shock waves are in fact compression waves caused by the expansion of magma gases and the projection of ejecta during explosions. These waves can be felt but in most cases they do not present any danger. However, material damage (broken windows and cracks in walls) and human damage (ear and organ damage) have been observed several tens to hundreds of kilometers from the source volcano during major explosions such as those at Tambora in 1815 and Krakatau in 1883 (Indonesia). The paroxysmal explosion of the morning of August 27, 1883, is to this day the loudest sound ever heard by humans in history. The wave traveled seven times around the globe and the explosion was heard as far away as Rodrigues Island in the Indian Ocean, 4,700 km from Krakatau (Strachey 1888).

A phase coupling phenomenon between the atmospheric pressure wave and the ocean surface can occur under very specific conditions (Ewing and Press 1955). The transfer of energy between air and water occurs only in large ocean basins. This phenomenon would be at the origin of the tsunami recorded by the tide gauges of the whole world during the paroxysmal explosions of Krakatau on August 27, 1883 (Pelinovsky et al. 2005) and Hunga Ha'apai Tonga on January 15, 2022.

2.2.6. *Pyroclastic density currents*

Pyroclastic density currents are generated during explosive volcanic eruptions or by gravity collapse of a lava dome (see Figure 2.15). They consist of hot mixtures of gases and solid particles, are denser than the surrounding air and consequently propagate laterally under gravity (Brown and Andrews 2015; Dufek et al. 2015). The gas and solid phases result primarily from magma fragmentation and may be enriched by ambient air ingested by the stream as well as by rock fragments detached from various parts of the volcanic edifice. The temperature of the gas-particle mixture is ~100°C–600°C and depends in particular on the quantity of air ingested.

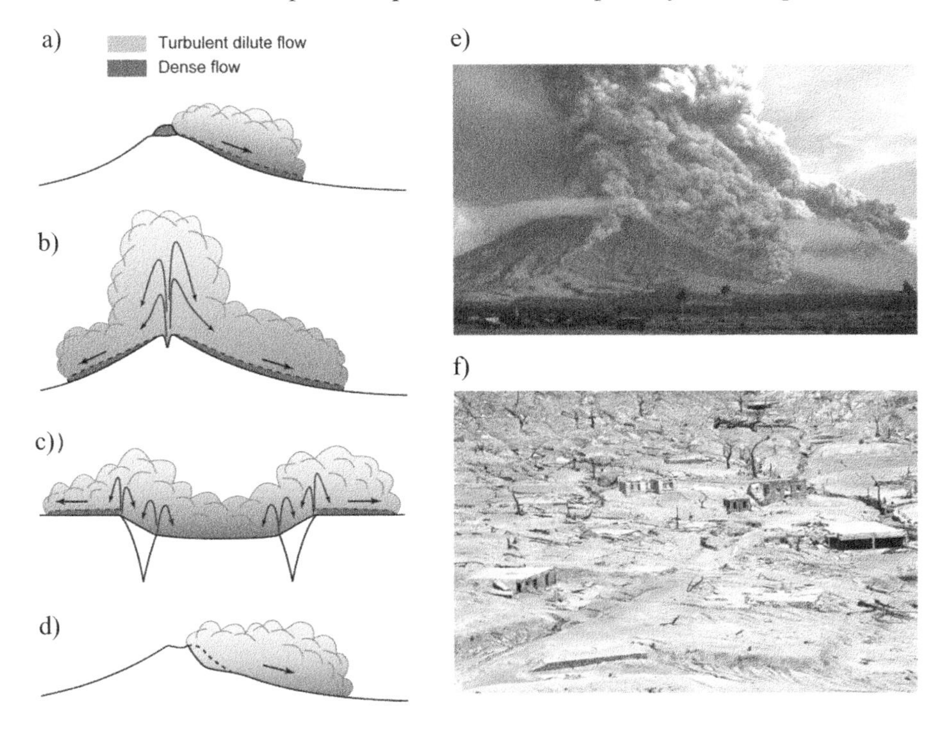

Figure 2.15. *Mechanisms of formation of pyroclastic density currents*

COMMENT ON FIGURE 2.15.– *a) Gravity collapse of a lava dome (Merapi 2006, Soufrière Hills 1995). b) Gravity collapse of an eruptive column (Vesuvius 79, Pinatubo 1991); note that a similar mechanism occurs in phreatomagmatic eruptions where only dilute flows are generated. c) Caldera formation (Tambora 1815, Toba 75 ka). d) Directed explosion caused by rapid decompression of a (crypto) lava dome (Mount St. Helens 1980). A dense basal flow is most likely to form in cases (a–c). e) Pyroclastic flows formed by collapse of an eruptive column at Mayon volcano, Philippines, 1984 (photo: C. Newhall). f) Inhabited area destroyed by pyroclastic flows at Soufriere Hills volcano, Montserrat, 2010 (Cole et al. 2015).*

Most of the physical parameters of pyroclastic density streams are characterized by several orders of magnitude (see Table 2.2). Particle (pyroclast) sizes range from micrometer ash to fragments typically on the order of centimeters and in some cases reaching several meters in size. The flow units within the streams have volumes of the order of 10^4–10^8 m^3 and their accumulation during sustained eruptions, as in the case of the collapse of a Plinian column or the formation of a caldera (see Figures 2.15(b) and 2.15(c)), leads to the formation of deposits with volumes of up to ~1,000 km^3 that are called *ignimbrites* (see Figure 2.16).

Parameters	Typical values
Particles	
Size	10^{-6}–1 m
Density	500–2,500 kg/m^3
Volume fraction	0.01%–50%
Flows	
Temperature	100°C–600°C
Density	1–1,500 kg/m^3
Velocity	10–150 m/s
Thickness	1–1,000 m
Travel distance	1–200 km

Table 2.2. *Main physical parameters of pyroclastic density currents and orders of magnitude*

2.2.6.1. *Transport mechanisms*

The propagation velocity of currents is typically ~10–50 m/s when driven solely by gravity and can reach 100–150 m/s in the case of directed explosions that result from the rapid decompression of a magma body (Mount St. Helens, 1980) (Figure 2.15(d)). Most currents are capable of propagating along slopes of a few degrees over distances that can reach 150–200 km in the case of some large calderic eruptions (>100–1,000 km^3, see Figure 2.15) (Streck and Grunder 1995; Wilson et al. 1995; Roche et al. 2016).

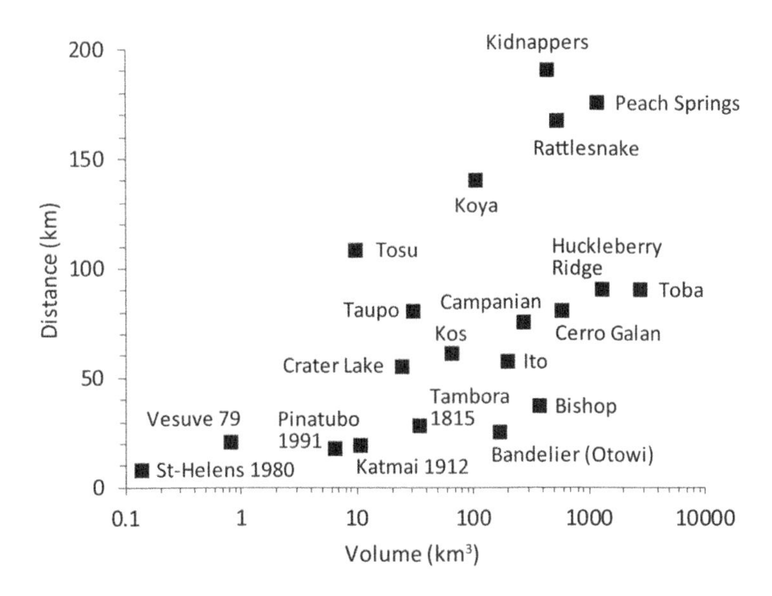

Figure 2.16. *Maximum travel distance of pyroclastic density currents formed by eruptive column collapse or during caldera eruptions (see Figure 2.15(b–c)) as a function of the volume of their deposit (ignimbrite) that results from the accumulation of flow units. Calderas are formed by eruptions with volumes greater than ~10 km^3*

The long travel distances, large volumes and high temperatures of pyroclastic density currents rank these volcanic hazards among the most devastating and deadly (see Figure 2.15(f)). To date, they have caused the deaths of more than 90,000 people, one-third of the losses recorded since 1600 (Auker et al. 2013). Pyroclastic density flows generated by the successive collapses of the Mount Pelée dome (Martinique) in May and August 1902 are largely responsible for the 28,800 recorded casualties (see Figure 2.17).

Figure 2.17. *Bourdon of the cathedral of St. Pierre de la Martinique, deformed by the heat of the pyroclastic flow of May 8, 1902 (Franck Perret museum, St. Pierre de la Martinique) (photograph: Raphaël Paris)*

During calderic eruptions, magma extraction leads to lowering the pressure in the magma chamber until the roof of the magma chamber collapses, creating a depression (the caldera) ~100–4,000 m deep and typically ~5–50 km in size at the surface (see Figure 2.15(c)). Several historical examples (Santorini ~1600 BCE, Tambora 1815, Krakatau 1883 and Pinatubo 1991) show that pyroclastic density currents generated during such eruptions have had severe consequences for populations and infrastructure on a regional scale. In the marine domain, dense flows may have entered the water and triggered tsunamis that washed ashore (Krakatau 1883 (Self and Rampino 1981); see section 2.3.3). The largest caldera-forming eruptions, called supereruptions when the volume of deposits is greater than ~1,000 km^3 (see Figure 2.16), have a recurrence on the order of 10,000–100,000 years. They have long-lasting environmental and climatic consequences on the scale of at least one hemisphere because of the surface area of the devastated areas and the dispersal of considerable amounts of ash and gas into the atmosphere (Self 2006; Miller and Wark 2008). The Toba eruption (~75 ka, ~2,800 km^3) is one such example (Costa et al. 2014).

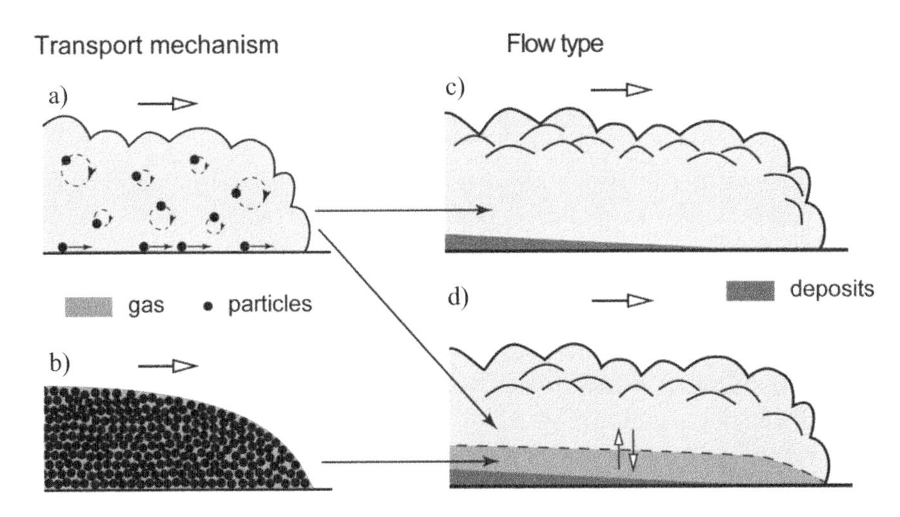

Figure 2.18. *Mechanisms of solid particle transport in gas-particle gravity flows and types of pyroclastic density currents. For a color version of this figure, see www.iste.co.uk/lenat/hazards.zip*

COMMENT ON FIGURE 2.18.– *a) Dilute flow with a turbulent suspension overlying a bottom load. b) Dense granular flow. c) Pyroclastic surge. d) Pyroclastic flow overlying a surge; the two parts of the pyroclastic flow can exchange material (vertical arrows), and their transition (dotted line) can be gradual or abrupt. In (c) and (d), the sedimentation of particles leads to the formation of a deposit whose size increases with time.*

The propagation of pyroclastic density currents is controlled by two solid particle transport mechanisms that can operate simultaneously in separate parts (see Figure 2.18). The first mechanism is the transport of pyroclasts in the form of a dense granular flow with a particle volume concentration of up to ~50%, giving the mixture a density on the order of ~10^3 kg/m^3. This flow forms the basal part of many pyroclastic density streams (called *pyroclastic flow*), which is topped by a turbulent dilute suspension (called *pyroclastic surge*). Interstitial gas has a negligible effect on the propagation of the granular mass when the average particle size is relatively large, as in the case of slag flows. These flows can only propagate on steep slopes close to the angle of friction of the natural material. On the contrary, the differential movement between gas and fine ash particles in high proportion, as in the case of pumice/block and ash flows, can lead to the emergence of a pore fluid pressure that considerably reduces the interactions between the particles and thus lowers the energy dissipation of the flow. This mechanism may be one cause of propagation over distances greater than 100–150 km, even when the topography is subhorizontal (Peach Spring tuff (Roche et al. 2016); Rattlesnake tuff (Streck and Grunder 1995)).

The second mechanism is the transport of pyroclasts by coupling with turbulent gas (*pyroclastic surge*). It operates in mixtures with low particle volume concentration, typically less than ~1%, and consequently with a density of the order of a few kg/m^3. The particle size range is smaller than in dense granular flows, with pyroclasts typically smaller than 0.1–1 cm. These turbulent mixtures overcome dense flows, but they can also individually form fully dilute flows, due to their formation mechanism (e.g. phreatomagmatic eruptions) or when they segregate from the dense basal flow when the latter encounters a sharp break in slope or abruptly changes propagation direction due to topography (see Figure 2.19).

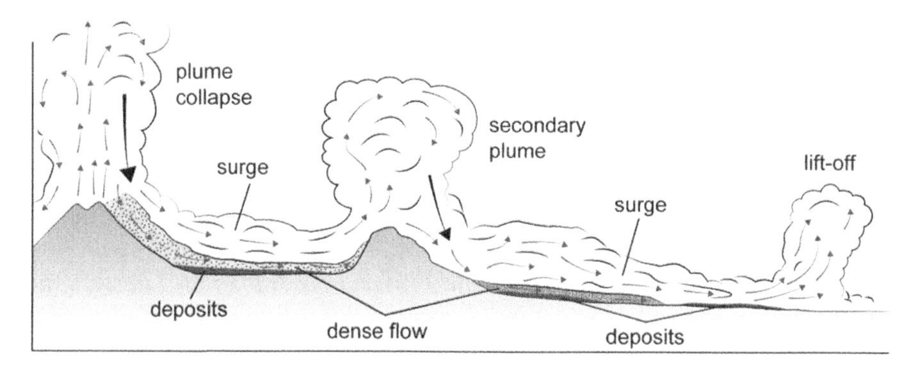

Figure 2.19. *Effect of topography on a pyroclastic density current generated by the collapse of an eruptive plume*

COMMENT ON FIGURE 2.19.– *The surge may separate from the dense flow when it encounters a topographic barrier and then concentrate to form a secondary flow that may have a dense basal component. The surge ingests ambient air at all stages and lift-off occurs if its density becomes lower than that of the atmosphere. Note the formation of deposits at the base of the dense flows and the surge in the distal parts. From (Wright et al. 2016).*

The detached surge can then concentrate and form a secondary dense basal flow (Druitt et al. 2002). Fully diluted flows consist of two distinct parts, a suspended load and a relatively fine, weakly turbulent bed load, which results from the decrease in velocity of the near-substrate flow and in which particles are transported by traction and saltation. Turbulent flows can ingest large amounts of ambient air. The air is then heated and expands, and consequently the density of the two-phase mixture decreases. Thus, some surge may become less dense than the atmosphere and can no longer propagate laterally. The gas-particle mixture then rises vertically as a convective ash plume (lift-off phenomenon, see Figure 2.4). During the Krakatau eruption in 1883 (Indonesia), a surge from the collapse of the Plinian column traveled more than 60 km across the waters of the Sunda Strait, burning more than a thousand people on the coast of Sumatra (Verbeek 1886).

2.2.6.2. *Related hazards and prevention*

As mentioned above, pyroclastic flows are among the most dangerous and deadly volcanic phenomena. Their speed and mass give them great kinetic energy. In addition to this, they have a high temperature and a toxic gas content. The consequences are always important:

– filling in of the topography;

– burying and/or destruction of everything in their path;

– combustion, burning of vegetation and buildings;

– burning and suffocation of living things (Jenkins et al. 2015).

There is no way to divert or contain pyroclastic currents. The only possible prevention is the mapping of exposed areas which is based on several complementary approaches. The first is geological. The study of the deposition of flows on a given volcano makes it possible to quantify and map the hazard on the basis of past or recent activity (Orsi et al. 2004). This approach can be enriched by numerical simulations that allow better defining of the potential impact for different eruptive scenarios. Different types of models have been developed to simulate these events. We will distinguish:

– simulations that consider the gas-particle mixture as a single-phase body and take into account a simple rheology, which are inexpensive in terms of computational time and therefore particularly useful for quickly identifying potential impact zones during an eruptive crisis (Kelfoun et al. 2009) or for defining probabilistic hazard maps based on several thousand simulations (Neri et al. 2015);

– multiphase flow simulations that take into account the behavior of pyroclasts and the gas phase as well as their coupling, which are costly in terms of computation time and are used not only to predict the impacted areas in the long term but also to study the physical mechanisms of the flows and the formation of their deposits (Ongaro et al. 2012). The hazard assessment is then conducted according to a deterministic approach based on scenarios referring to well-documented eruptions (Merapi; Thouret et al. 2000) or a probabilistic approach based on statistical models (Vesuvius and Phlegrean Fields; Sandri et al. 2018).

Finally, during a crisis, the monitoring means will allow for anticipation of the occurrence of flows and recording their progression. It can be noted (see Volume 2, Chapter 1) that seismic and acoustic networks can detect and monitor in real time the propagation of pyroclastic flows and also lahars. In exceptional cases, the design of shelters can be considered occasionally (protection of critical installations). Moreover, Louis-Auguste Cyparis survived the eruption of Mount Pelée in 1902 in a thick-walled, semi-buried dungeon whose small openings, opposite the volcano, were closed.

2.3. Indirect volcanic hazards

2.3.1. *Lahars and associated flows*

Water and sediment flows on volcanoes are among the most lethal and destructive volcanic hazards. A term of Indonesian origin, a lahar refers to a mixture of debris and water initiated on the flanks of a volcano that travels at average velocities of 3–10 m/s as it channels through the drainage system (Vallance and Iverson 2015). Since the year 1600, a quarter of the total number of casualties related to volcanic activity have been caused by lahars. This figure places them as the third most deadly volcanic hazard after pyroclastic flows and tsunamis (see Table 2.3).

2.3.1.1. *Types of lahars*

Because lahars are water-saturated, fluid-solid interactions influence their behavior and distinguish them from debris avalanches and jökulhlaups (torrential-type mass flow related to breakup following subglacial eruption and reservoir rupture within an ice sheet). Lahars are part of a continuum of water-rich flows, but sediment concentration, particle size distribution and density help distinguish the following two categories (Hungr and Jakob 2005; Iverson 2014):

– *Hyperconcentrated flows* are two phased, with a particle concentration of 20%–60% by volume and a density of 1,300–1,800 kg/m^3 between those of floods and debris flows. They are dilute, turbulent and their sandy-gravelly deposits are less heterometric and better sorted than those of debris flows;

– *Debris flows* are mixtures of debris and water that flow by gravity in a pulsating pattern. The solids concentration is higher (≥60% volume) than that of hyperconcentrated flows, the density is between 1,800 and 2,400 kg/m^3 and the deposits are massive, heterometric and very poorly sorted. The range of particles extends from sand to blocks of several meters, while the proportion of clays and silts remains low in general: from 3%, the lahars are said to be *cohesive*.

Volcano	Date	Volume 10^6 m^3	Distance *runout* km	Victims/affected, displaced	Damage to habitat, infrastructure, issues
Galungung	1822	>0.11	10–30	4,000/thousands	114 villages
Cotopaxi	1877	370	300	1,000/several thousands	Very extensive burial
Ruapehu	1953	1.9	60	151/>150	Roads, bridges, railroads
Mount St. Helens	1980	50.10^6	100	56/hundreds	Extensive damage: bridges, roads
Nevado del Ruiz	1985	90	104	23,030/4,420 injured	5,092 houses
Pinatubo	1991–1993	30–40	45	957/249,370 affected; 5,000 displaced	112,236 houses
Casita	1998	3.1	30	2,513 and 1,000 missing	2 buried cities
Sarno (Vésuve)	1998	1.42	0.9–2.1	150–161/>50 evacuated	Several dozen buildings and vehicles
Mayon	2006	18.8	8-14	1,266/>10,000 displaced	6 cities, 81,000 houses, roads, plantations, dams
Chaitén	2008	1–3	10 + delta	Many/>2,000 displaced	Inundated city, roads, bridges, port
Merapi	2011–2012	>5	20	3 and 15 injured/3,000 homeless	860 houses (215 devastated), 70 ha of cultivated land buried

Table 2.3. *Examples of historical lahars and associated impacts*

The processes of lahar initiation depend on the local conditions of the watershed. Favorable factors are loose lithology, poorly vegetated soils and a shallow water table. Water input can come from disturbed lakes or rivers, snow or ice melt, but most of the time it is long-lasting and intense rainfall (typically triggering from 20 to 30 mm/h) that causes lahars. On the slopes, the initial water-debris mixture deforms and liquefies as the pore pressure increases. By eroding flow channels and incorporating sediment, a lahar can increase in volume 5–10 times, acquiring a maximum velocity of 20–30 m/s and a maximum discharge of 10 times greater than that of floods in a watershed (Costa 1988).

There are three categories of lahars:

– *Syn-eruptive or primary lahars* are triggered by the overflow of crater lakes, subglacial eruptions, snowmelt and the eruption of pyroclastic flows or landslides into rivers or lakes. For example, large primary lahars (9×10^7 m^3), triggered by melting snow as pyroclastic flows pass over the Nevado del Ruiz ice cap, caused the death of 23,000 people in 1985 in Armero, located 56 km from the crater (Pierson et al. 1990);

– *Post-eruptive lahars* result from the remobilization of tephra by intense rainfall after large eruptions. They occur for several years after an eruption. The considerable damage they cause is due to the diversity and repetition of the physical impacts;

– Unrelated to eruptions, *secondary lahars* are triggered on the slopes and deposits of dormant devices subjected to intense meteorological events (e.g. Hurricane Mitch in 1988 on Casita, Nicaragua (Scott et al. 2005)). Some secondary lahars result from remobilization of ash far from its source, for example, about 100 km east of Vesuvius at Sarno in 1998 (Zanchetta et al. 2004), or from the transformation of slides on inactive volcanoes.

2.3.1.2. *Lahars emplacement and associated deposits*

The different sedimentary facies of lahars (see Figure 2.20) result from a continuum of flow types, depending on variations in volume, flow rate, sediment load and the distance reached beyond the break in slope between the cone and its foothills (called runout).

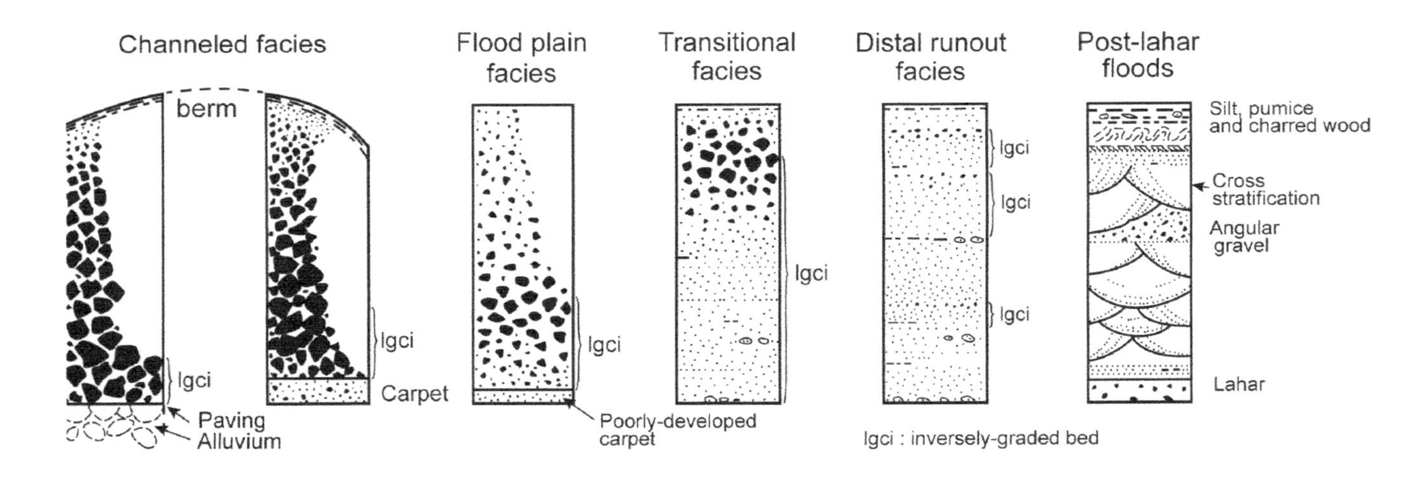

Figure 2.20. *Typical examples of lahar deposits and facies and associated flows. Typology according to the morphology of the valley through which the lahars flow and the succession of flows from upstream to downstream and over time (Scott 1988). Berm: a transitional, oblong, "whale-back" shape caused by coarse deposition in a river. Runout, term explained in the text. Lgci: reverse granular grading bed. Adapted from Scott (1988)*

Figure 2.21. *Boulder-rich lahar front, 4 m high and 18 m wide, advancing at 4.5 m/s in the Lengkong River at Semeru volcano in February 2002 (photo: Franck Lavigne)*

Lahars generally have broad, thick fronts (see Figure 2.21), with coarse material exerting high basal friction. They grow downstream of the channel, pushed back by finer debris, less abundant and less slowed by their lower friction on the channel bed. In the case of cohesive lahars (rich in silts and clays), a sediment plug propagates in the channel "en masse" (sudden slowing down of the whole mass of the deposit, without stratification) or by progressive aggradation (a rapid succession of more or less stratified deposits). Hyperconcentrated flows have non-Newtonian behavior, but because their internal cohesion is less than that of debris flows, these flows are dominated by friction and interactions between particles (Manville et al. 2009). Hyperconcentrated flows therefore exert less dynamic pressure on the structure and infrastructure compared to debris flows.

2.3.1.3. *Impact of lahars*

The physical impacts of lahars on habitat, infrastructure and networks have been analyzed in detail from recent case studies (Jenkins et al. 2015; Thouret et al. 2020) and hazard reduction strategies have been proposed (Pierson et al. 2014). Lahar impact forces are often restricted to dynamic pressure and inundation depth that induce physical actions (Wilson et al. 2014, 2017), but other direct and indirect processes must also be considered (see Table 2.4).

Figure 2.22. *Lahar impacts in Kali Putih Valley, Merapi volcano in 2011 following the 2010 Merapi eruption (Jenkins et al. 2015): a) flooding marks on walls; b) destroyed buildings; c) dam rupture. Adapted from Jenkins et al. (2015). For a color version of this figure, see www.iste.co.uk/lenat/hazards.zip*

COMMENT ON FIGURE 2.22.– *a) Interior of a large building in Sirahan showing the lines (red dashes) of inundation left by the lahar and its deposition (photo S. Jenkins). b) Destroyed building revealing ruptured reinforced concrete masonry walls and remnants of deposits inside; direction of flow from right to left. c) Annotation of lahar damage on Gejugan I dam upstream of K. Putih in 2010 (Jenkins et al. 2015).*

Forces and actions		Short-term induced processes	Long-term consequences
Pressure	Horizontal dynamics	Wall deformation (differential thickness of the deposit between the interior and exterior of the building)	Brittle deformation, fractures, breakage
		Disintegration. Lateral displacement, collapse Displacement of walls and roof via fractures	After the collapse of the building, material remobilized and transported
		Material deformation resulting from water infiltration	Metal corrosion
	Vertical dynamics	The roofs absorb the deformation up to the "plastic" threshold Blocking of doors and windows preventing evacuation	Ductile deformation; disintegration Turbulence of the flow if it persists
		Fractures in roofs leading to disintegration. Fall of roof and slabs	Brittle deformation. Vertical displacement, fractures, collapse
	Hydrostatics	Moisture rise by capillary action. The material may exceed the plasticity threshold	Deformation of load-bearing walls and propagation to other building elements
		Fracture propagation through load-bearing walls leads to collapse	Prolonged soaking through walls from saturated soil weakens foundations and walls

Forces and actions	Short-term induced processes	Long-term consequences
Flooding and deposits	Burial blocks the doors Infiltration makes the structure less resistant and can corrode it Can remobilize the deposit surrounding the building	Water soaks in, weakens and deteriorates the structure Corrosion can destabilize metal structures Modification of the topography
Erosion	Lateral erosion and scouring of the substrate under the building and deformation induced by its mass	Deformation, fractures, displacement and collapse
	Removal of protruding elements from structures	Destabilization and displacement of the structure, or even transport out of the area
Buoyancy	The buoyancy force lifts the structure and objects	Lifting and flotation by capillarity. Deformation and object collisions
Boulder impacts	Displacement or collapse of walls due to impacts	Fracture propagation due to high energy boulders
	Boulder impacts create cavities in buildings	Breakdown and collapse
Temperature	Combustion of flammable material. Resistance of certain materials affected	Slow combustion by infiltration of hot deposits. Chemical and biological reactions weakening the mortar or binder
Contamination	From the outside if overflow or avulsion	Acid contamination, fuel and waste water

Table 2.4. *Damages caused by lahars to buildings*

The specific characteristics of each lahar determine its impact capacity. During a prolonged pulse-marked flow, the transition and succession of hyperconcentrated flows and debris flows from upstream to downstream of the channel result in alternating burial and incision phases during each pulse (Manville et al. 2009; Doyle et al. 2010, 2011). The extent of the inundated area depends not only on the mobility of the lahar and its runout distance, but also on the morphology of the channel. Lateral levees and steep channel slope promote high velocities and runout distances, while on the plains, channel widening reduces them. Lahars overtake the main channel when the channel's transport capacity is reduced and when the slope and sinuosity of the riverbed increases (Solikhin et al. 2015). The potentially affected area then increases beyond the main valleys by the fact that lahars, having overtopped, move through secondary valleys where populations and infrastructure usually not at risk reside. The process of avulsion refers to these lahars, or pyroclastic flows, which pass through secondary valleys. This avulsion phenomenon is common on the foothills and low-lying plains surrounding active volcanoes with a very dense drainage network (e.g. Pinatubo in the Philippines and Merapi in Indonesia).

2.3.1.4. *Monitoring of lahars*

A wide range of methods can be implemented to delineate exposed areas (see Chapter 3), predict damage, measure flow behavior in real time, detect their initiation and displacement, and limit their impact with artificial protective structures. Different measurement methods and protocols can be used for lahar monitoring and characterization (Itakura et al. 2005). Among the main ones are:

– sampling at regular time intervals to analyze particle distribution, density and sediment concentration, as well as temperature and water chemistry to trace their origin;

– cameras recording flow velocity, block transport and surface hydraulic instabilities;

– radar or ultrasonic level meters to measure flow height;

– pore pressure sensors to deduce sediment concentration.

Monitoring of lahar occurrence and flow is primarily based on precipitation monitoring and specifically adapted seismic and/or acoustic monitoring. The amount and rate of precipitation can provide clues to the probability of lahar initiation based on volcano morphology and the presence of mobilizable materials (Lavigne et al. 2000; Lavigne and Thouret 2000). The waves resulting from the coupling of the flow with the ground are most intense in the range above 10 Hz, which is the highest frequency of most conventional seismic signals. Nevertheless, conventional seismic arrays can be used to detect and track flows, as they are used to record various surface signals. Specific arrays have been developed for flow monitoring, for example, at Merapi, Ruapehu (Cole et al. 2009) or Colima. The sensors are geophones adapted to signals between 10 and 300 Hz. This type of array, developed by the U.S. Geological Survey, is called Acoustic Flow Monitors (AFM). Beyond the simple detection of lahars, the frequency analysis of signals can provide information on the composition and dynamics of flows.

2.3.2. *Prevention of lahars*

The prevention of lahar-related hazards is based on:

– mapping of potentially impacted areas and informing the population (see Figures 2.23 and 3.4);

– modeling of lahar flow;

– the development of valleys through different types of structures.

Numerical models help to understand the dynamics and behavior of lahars (Turnbull et al. 2015), but the variability of rheology during flow remains difficult to define and incorporate into models. Many numerical models have been developed over the past 20 years:

– statistical codes, such as LAHARZ-py (Schilling 2014), and propagation models are used to predict runout distances;

– 2D models based on Saint-Venant equations, such as TITAN2D (Pitman and Long 2005) and VOLCFLOW (Kelfoun and Druitt 2005);

– 3D models such as the Smooth Hydrodynamics Particle Lagrangian code (Pastor et al. 2009), which allows engineers to track the behavior of flow particles, estimate velocity and dynamic pressure, and assess impacts on structures using highly accurate topographic models (Mead et al. 2016).

The presence of a lake in a crater is an element that favors the formation of lahars. Two emblematic cases are well known: Ruapehu in New Zealand and Kelud (or Kelut) in Indonesia. In Ruapehu, the lake and its moraine dam are monitored by automatic cameras and by instruments measuring the chemical and thermal properties of the lake, and the downstream infrastructure is protected. On the other hand, in Kelud, successive tunnels have been dug since 1919 to drain the lake. This system has proven to be very effective, but as the tunnels are systematically blocked or damaged after each eruption, it was necessary to clear or recreate them regularly until 2014. Since 2014, the dome fills almost completely the depression formerly occupied by the lake.

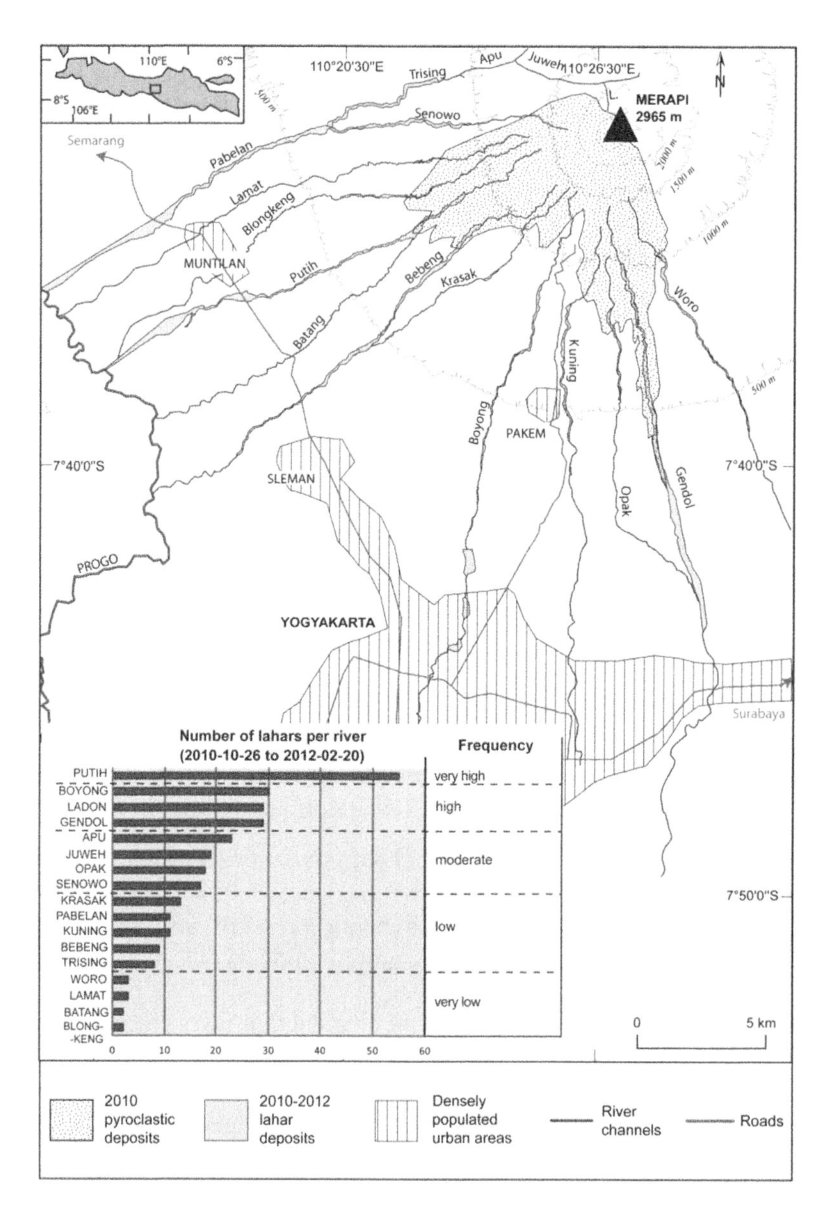

Figure 2.23. *Map showing the drainage network and pyroclastic (gray) and lahar (yellow) deposits from the 2010 eruption around Merapi (De Bélizal 2012). The inserted diagram shows the number of lahars for each of the 17 rivers for 2 years after the 2010 eruption. Urban and densely populated areas are dashed. For a color version of this figure, see www.iste.co.uk/lenat/hazards.zip*

Downstream of their trigger zone, protective structures against lahars and debris flows have long been developed in Japan and then exported to many sites around the world. This type of structure is known as a Sabo in Japanese. These structures aim to break up the flows in the valleys by reducing their load and channeling them (see Figure 2.24).

Figure 2.24. *Retention and filtration dams for lahars (Sabo dam) built on the eastern flank of the Unzen volcano in Japan, following the 1990–1995 eruption (photo: Raphaël Paris)*

2.3.3. *Landslides and debris avalanches*

An entire flank of a volcano may slip, break up and then flow as a mixture of crushed rock with particularly fluid behavior, devastating areas downstream. Volcanic edifices are indeed relatively unstable reliefs. This flank instability is materialized by a whole range of phenomena from simple rockfalls and rock wall collapses (typically $<10^6$ m^3) to major instabilities such as debris avalanches ($1–10^3$ km^3). The causes of this instability are both endogenous, that is, related to the dynamics of the volcano, and exogenous (external factors). The main endogenous causes are:

– rapid rates of accumulation of volcanic products (often on steep slopes);

– the presence of structural discontinuities in the edifice (faults, lithological discontinuities);

– intrusive dynamics (upwelling of veins, magma veins);

– seismicity (often linked to intrusions);

– hydrothermal alteration (Siebert 1984; Keating and McGuire 2000; van Wyk de Vries and Delcamp 2015).

Exogenous causes of instability include heavy rainfall events, regional tectonics and associated seismicity, or eustatic variations. Exogenous causes often intervene in the triggering of moderate volume landslides (e.g. landslides after a cyclone) or as an aggravating factor of a destabilization of endogenous origin. If the causes of instability are well known, identifying the triggering factor or the combination of factors is sometimes problematic. As an example, the 1792 Mayu-yama peripheral dome debris avalanche at Unzen (340×10^6 m^3 (Michiue et al. 1999)) was largely driven by hydrothermal fluid pressure, but the triggering of the avalanche coincided with the strongest earthquake of the eruptive crisis.

In terms of recurrence, landslides such as those that affected Stromboli in 2002 (17×10^6 m^3 and 5×10^6 m^3 (Bonaccorso et al. 2003)) are relatively frequent, on the order of several events per decade. The volumes involved in the 2002 Stromboli slides are two orders of magnitude smaller than the largest historical debris avalanches, such as Mount St. Helens in 1980 (2.8 km^3 (Voight et al. 1981)), Ritter Island in 1888 (5 km^3 (Cooke 1981)) and Oshima-Oshima in 1741 (2.4 km^3 (Satake and Kato 2001)). The largest debris avalanche to be directly observed was from Mount St. Helens. After a few weeks of intense seismic activity and deformation of the northern flank of the volcano, a volume of 2.5 km^3 collapsed, devastating an area of more than 60 km^2 (Voight et al. 1981 and see Figure 2.25). This event made the volcanological community aware that phenomena of this magnitude could occur. Although much smaller in volume (<0.2 km^3), the debris avalanche from the Anak Krakatau cone (Indonesia) in December 2018 caused a deadly tsunami in the Sunda Strait (see Figure 2.26; Gouhier and Paris 2019; Walter et al. 2019).

There are several scenarios of volcanic debris avalanches (Siebert et al. 1987):

– because the edifice is destabilized by voluminous magma intrusions (e.g. crypto-dome), the debris avalanche often occurs during the eruptive paroxysm and in this case is associated with a lateral explosion and pyroclastic flows (e.g. Mount St. Helens in 1980 and Bezymianny in 1956);

– but the debris avalanche can also affect zones on the periphery of the active edifice, without the activity being necessarily explosive (zones weakened by hydrothermal alteration: e.g. Mayu-yama 1792);

– or even without surface volcanic activity, during a short phreatic paroxysm without precursory signs (e.g. Bandai-san 1888).

Figure 2.25. *Succession of shots showing the formation of the debris avalanche of Mount St. Helens on May 18, 1980 (photo: Gary Rosenquist)*

Debris avalanches are particularly mobile and can travel several tens of kilometers and spread over areas of several hundred km^2 at speeds of more than 300 km/h (Kelfoun and Druitt 2005). Geological observations indicate that mobility depends on the volume collapsed: the larger the volume, the greater the distance reached (see Figure 2.27). This extreme mobility can also be observed in the morphology of the deposits that literally cover the landscape (see Figure 2.28): even if their thickness reaches several tens of meters, it remains low compared to their extension. These heterogeneous deposits have an irregular surface morphology (hummocky relief and megablocks several tens to hundreds of meters wide) and their structure includes faults, folds and lateral (levees) or frontal scarps (Siebert et al. 1987; Glicken 1996; Shea and van Wyk de Vries 2008; van Wyk de Vries and Delcamp 2015).

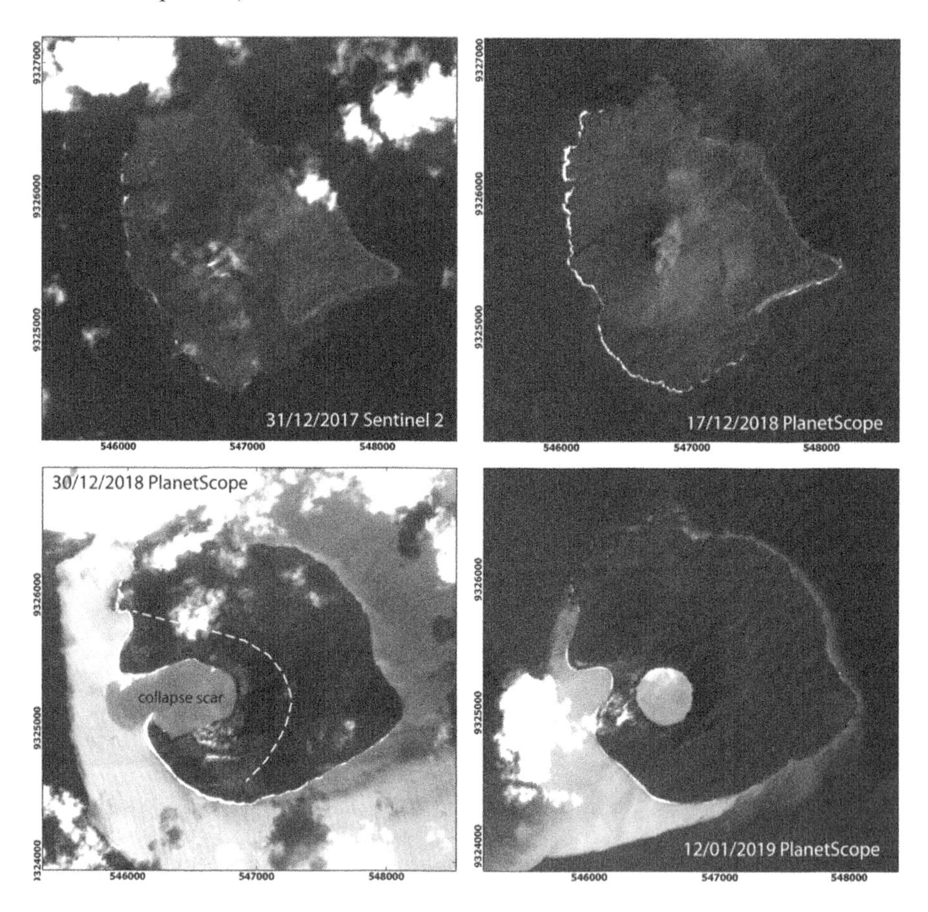

Figure 2.26. *Evolution of the Anak Krakatau volcano between December 2017 and January 2019 illustrated by satellite images (Sentinel-2 and PlanetScope). The December 22, 2018 collapse swept more than 100 million m³ into the waters of the Sunda Strait, causing a deadly tsunami on the coasts of Java and Sumatra (Gouhier and Paris 2019). For a color version of this figure, see www.iste.co.uk/lenat/hazards.zip*

In terms of volume, thickness and surface area, debris avalanches destroy everything they encounter and no human construction can withstand them. They have the capacity to dam rivers, resulting in the formation of lakes and the risk of flooding in the event of dam failure (Hall et al. 1999).

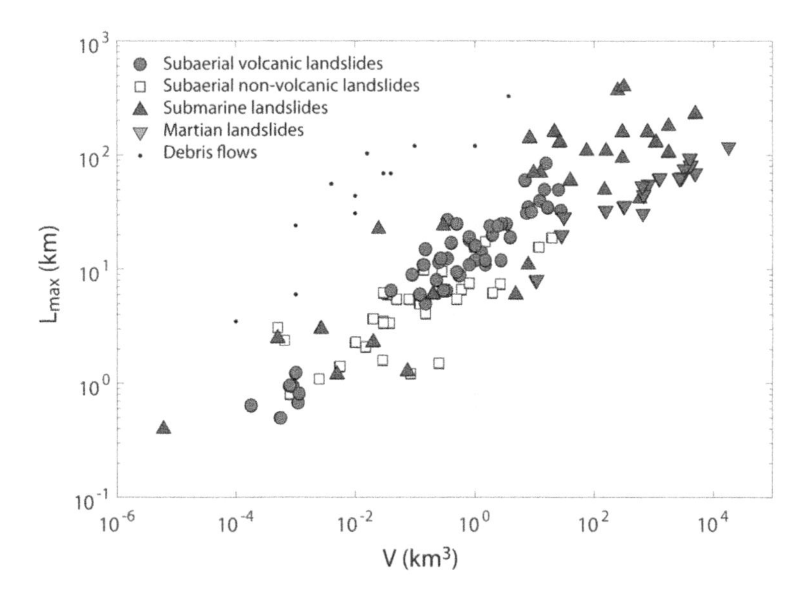

Figure 2.27. *Distance L reached by debris avalanches of volcanic origin as a function of their volume V compared to different types of non-volcanic landslides. Adapted from Legros (2002). For a color version of this figure, see www.iste.co.uk/lenat/hazards.zip*

Numerical modeling is a particularly effective tool for simulating debris avalanche emplacement (Cuomo 2020). It is able to reproduce the velocities, extensions, thicknesses and surface structures. The numerical simulations also illustrate the very high fluidity of these phenomena. For example (see Figure 2.29), a simulation of the Socompa volcano debris avalanche (Chile) shows that in order to reproduce the deposits, the avalanche must have traveled 30 km northward and accumulated on the northern edge of the topographic basin before returning toward the volcano over a distance of 15 km, remobilizing the deposits that were put in place a few tens of seconds earlier (Kelfoun and Druitt 2005). This behavior is reminiscent of water, but the avalanche consists exclusively of rock fragments (<10 µm and >10 m^3) and no observation of the deposits reveals the presence of water at the time of collapse. This "rock tsunami" formed a wave about 100 m thick moving at over 300 km/h.

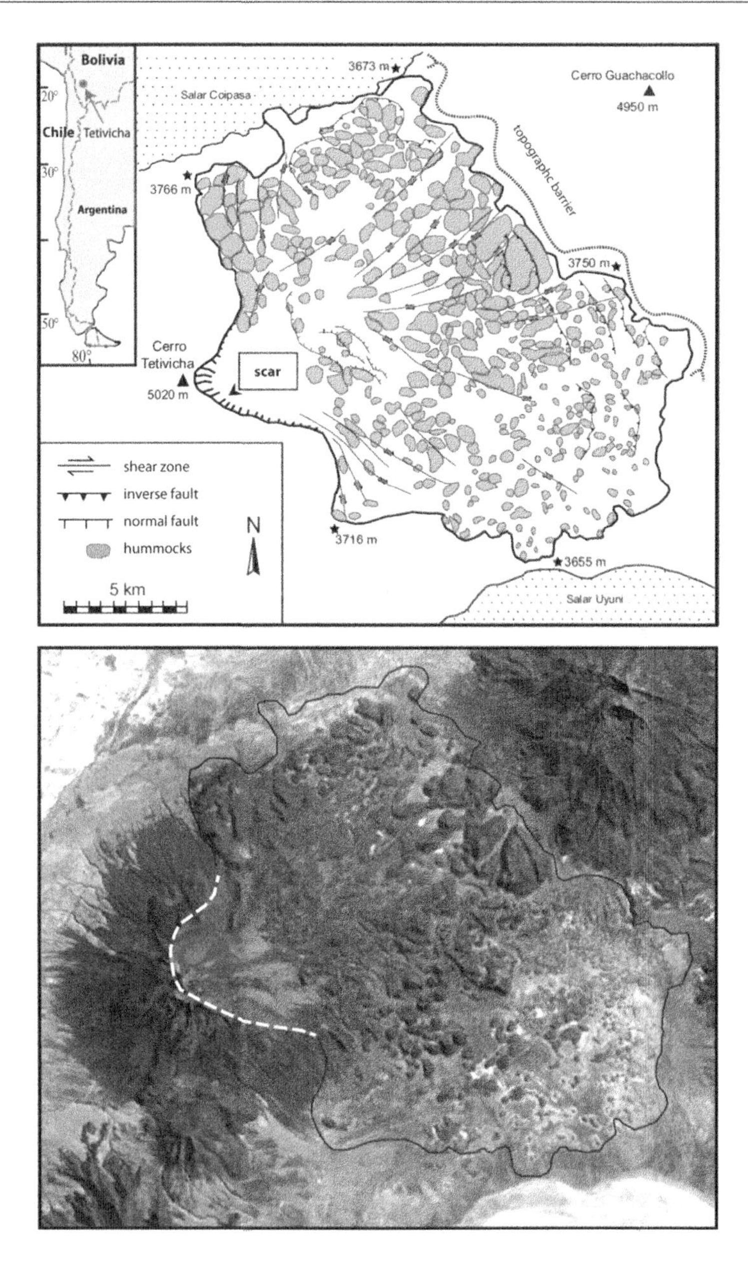

Figure 2.28. *The debris avalanche from Tetivicha volcano (Bolivia). The deposit is characterized by a hummocky morphology and structured into normal faults (proximal extensional zone), reverse faults (distal compressional zone) and longitudinal shear zones. Modified according to Shea and van Wyk de Vries (2008)*

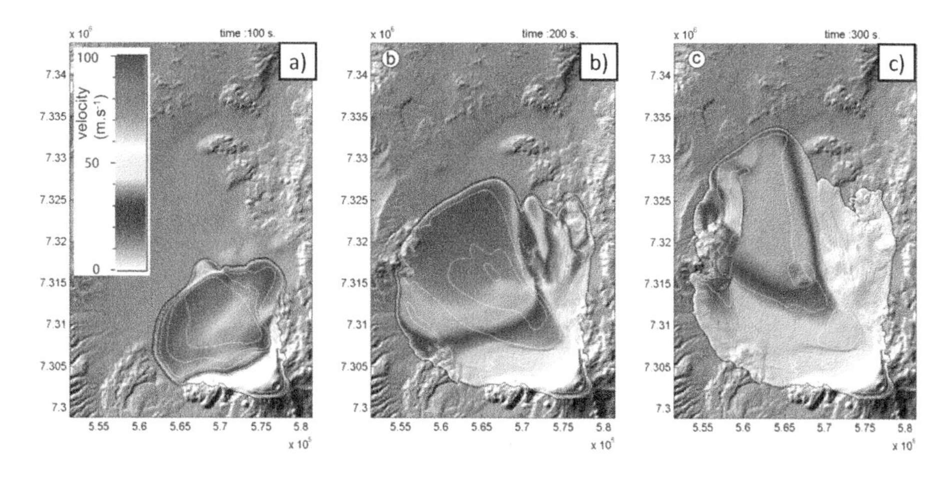

Figure 2.29. *Simulation of the Socompa debris avalanche emplacement, Chile (Kelfoun and Druitt 2005). For a color version of this figure, see www.iste.co.uk/lenat/hazards.zip*

COMMENT ON FIGURE 2.29.– *From left to right: evolution of the debris avalanche over time (100 s, 200 s and 300 s after destabilization). a) The debris avalanche initially flows northward; b) then it is influenced by the relief on which it is emplaced; (c) it flows back. This simulation illustrates the extreme fluidity of debris avalanches, yet they consist of finely ground rock (< mm) and metric blocks.*

Even if it is possible to simulate precisely the setting up of debris avalanches, the physical behavior used in the models is not explained. One or more phenomena explain why the rocky mixture acquires an extreme fluidity. For some authors, the various volatiles present within the volcano (water tables and gases) could explain this behavior, but other mechanisms have been evoked such as fluidization, lubrication and dynamic disintegration (Legros 2002). This debate, which has been going on for several decades, is still unresolved.

Shield volcanoes and especially oceanic islands are also subject to instability phenomena. The landslide scars identified on land or at sea and the debris avalanche deposits observed on the seafloor involve gigantic volumes of the order of tens to hundreds of km^3 (Moore et al. 1989; Carracedo et al. 1999; Oehler et al. 2004). Due to a lack of instrumental data on these out-of-ordinary events, it is currently difficult to infer the mechanisms controlling these giant flank slides and to assess the risks associated with them, including the risk of tsunami (Paris et al. 2018).

During the major eruption of Piton de la Fournaise in 2007, the eastern flank of the volcano slids several tens of centimeters along a large detachment plane (Froger et al. 2015). This example clearly confirms the close links between intrusive activity, volcano structure and flank instability. Shield volcano instability can also occur in a more gradual way, as can be seen currently on the southern flank of Kilauea in Hawaii. The volcano is continuously monitored by a network of geodetic instruments (GPS, inclinometers and pressure sensors) that allows researchers to follow the evolution of the unstable zone. This type of slip is often described as incremental because it results from the combined activity of a network of listric normal faults cutting the southern flank of the volcano. The slip zone experiences phases of aseismic acceleration (e.g. November 2000 (Cervelli et al. 2002)), but it is also affected by earthquakes of magnitude M > 6 (e.g. the 1975 earthquake, Ms = 7.2 (Ando 1979); or the 2018 earthquake, Mw = 6.9).

2.3.4. *Tsunamis*

Although accounting for only 5% of recorded tsunamis over the past four centuries (about 20 events per century), tsunamis caused by volcanic eruptions and volcano slope failures are often deadly (Begét 2000; Paris 2015). The human toll is high, as tsunamis thus account for 20% of volcanic casualties, or 55,300 deaths for which five major events are responsible (Auker et al. 2013): Krakatau in 1883 (Indonesia, more than 30,000 victims due to tsunamis), Unzen in 1792 (Japan, 14,500 victims), Ritter Island in 1888 (Papua New Guinea, between 500 and 3,000 victims), Oshima-Oshima in 1741 (Japan, 2,000 victims) and Iliwerung in 1979 (Indonesia, 580 victims).

Yet, volcanic tsunamis are understudied compared to other volcanic hazards and are not sufficiently considered in volcanic risk assessment. Tsunami warning systems are mainly adapted to tectonic tsunamis. The tsunami generated by a landslide on Anak Krakatau (Indonesia) in December 2018, for example, hit coasts that were nevertheless equipped with sirens and evacuation routes, but, in the absence of an earthquake, no warning could be issued. The vast majority of volcanic tsunami victims are located less than 20 km from the volcano, which reduces the time available to issue a warning. Impacted coastlines are not necessarily threatened by other volcanic hazards and are therefore potentially out of scope in terms of monitoring (Paris et al. 2014a). Some of the instrumentation of volcanological observatories located near coastal active volcanoes could be adapted or dedicated to tsunamis. Only Stromboli has benefited from a land-sea monitoring arrangement since the

2002 eruption and tsunami. A tsunami warning was effectively triggered in 2007 (Bertolaso et al. 2009). Active submarine volcanoes, on the other hand, are hardly monitored at all. The challenge is not only instrumental but also involves increasing public awareness.

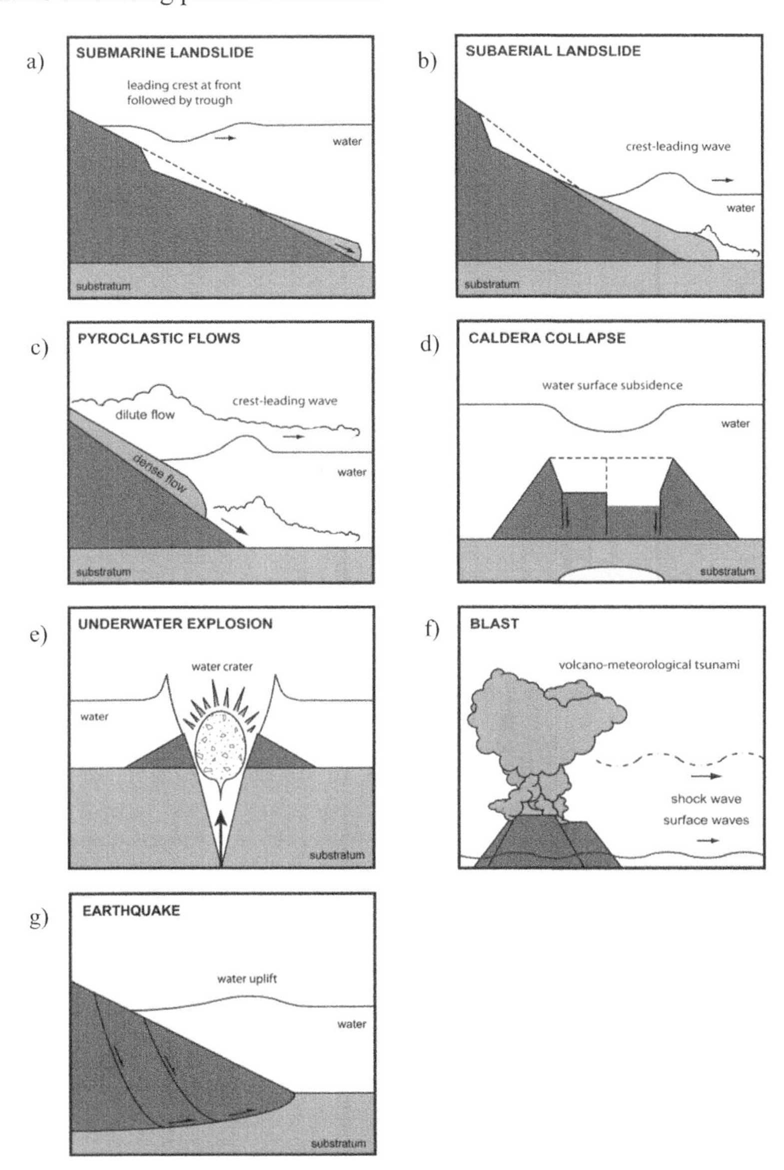

Figure 2.30. *Tsunami source mechanisms in a volcanic setting. Modified from Paris (2015)*

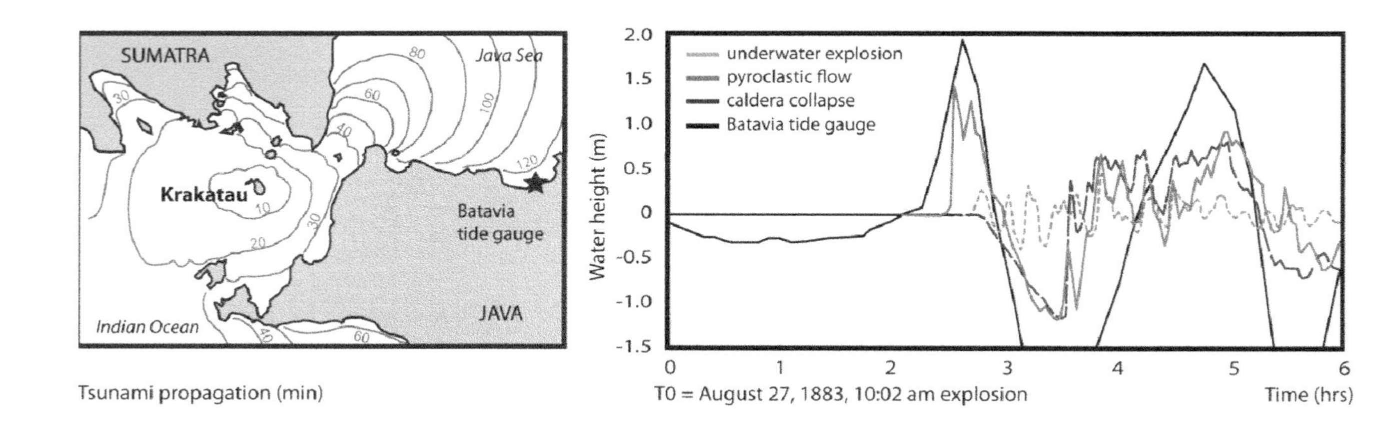

Figure 2.31. *Modeling of the tsunami generated during the volcanic explosion of August 27, 1883 at 10:02 a.m. Three types of sources are tested and compared with the Batavia (Jakarta) tide gauge: underwater explosion, pyroclastic flow and caldera collapse. Modified from Maeno and Imamura (2011). For a color version of this figure, see www.iste.co.uk/lenat/hazards.zip*

The source mechanisms for tsunamis are varied in volcanic contexts (see Figure 2.30): pyroclastic flow, flank collapse, submarine caldera collapses, subaqueous explosion, volcano-tectonic earthquakes, etc. The energy and volumes involved, as well as the characteristics of the waves generated, vary from one mechanism to another (Paris 2015). During major explosive eruptions, it is sometimes difficult to identify the source of a tsunami, especially in the case of an eruption leading to the formation of a submarine caldera (see the numerous publications on the Krakatau eruption in 1883, or on the Minoan eruption of Santorini). Another difficulty can also come from the frequency of tsunamis during a single eruption: at least 10 tsunamis were observed on the west coast of Java during the Krakatau eruption between August 26 and August 27, 1883 (Verbeek 1886).

The eruption of Krakatau volcano in August 1883 is among the most iconic eruptions in volcanology (Simkin and Fiske 1983). Collections of eyewitness accounts and post-eruption measurement campaigns (Verbeek 1886), as well as more recent studies (Paris et al. 2014b; Carey and Bursik 2015), have reconstructed the scenario of the events in detail. The Plinian-type eruption has remained famous for the tsunamis it caused, repeatedly ravaging the coasts of the Sunda Strait. The causes of these tsunamis have been debated at length (see Figure 2.31): pyroclastic flow spreading into the sea, collapse of the sides of the volcano or caldera, phreatomagmatic explosion? The major tsunami is caused by the setting of ignimbrite at sea during the eruptive paroxysm. Waves of 15 m reach altitudes of 40 m on the coasts of the Sunda Strait. The shock wave of the main explosion (morning of August 27, 1883) finally went several times around the earth and generated a last tsunami of very low amplitude ("volcano meteorological" tsunami) but which will be recorded by the tide gauges of the whole world.

In a review article, Paris (2015) identifies several problematic scenarios:

– a Bandai-type debris avalanche (related to a phreatic explosion) without precursors;

– a Mayu-yama-type debris avalanche (as at Unzen in 1792: see Figure 2.32) affecting a volcanic edifice at the margin of active vents and thus not covered by the monitoring network;

– an explosive lacustrine eruption that suddenly becomes tsunamigenic;

– any type of tsunami caused by the activity or instability of a submarine volcano.

Figure 2.32. *Map of the impact of the eruption of Unzen volcano and the tsunami caused by the collapse of the Mayu-yama peripheral dome in 1792 (drawing on cloth, Tokiwa Museum of Historical Materials, Honkoji Temple, Kyushu) (photograph: Raphaël Paris). For a color version of this figure, see www.iste.co.uk/lenat/hazards.zip*

Modeling volcanic tsunami scenarios to better assess and anticipate this potentially major hazard is a real challenge (see Figure 2.33). The numerical codes used vary according to the types of sources (explosion, flow, etc.) and the bathymetric context (Yavari-Ramshe and Ataie-Ashtiani 2016). Only multi-fluid models based on the Navier-Stokes equations can simulate complex tsunami sources such as landslides or pyroclastic flows and reproduce the interaction phenomena with coastal structures and inundation on land. But the computation time required can become a limiting factor. To simulate wave propagation, commonly used models solve the St. Venant equations (shallow water equations), a form derived from the Navier-Stokes equations for fluids with a horizontal component of velocity much higher than the vertical component ($L/h >> 20$). These equations can be used:

– in a linear form to model the propagation of a tsunami at the scale of an ocean basin;

– in a nonlinear form (integrating a friction term related to the water depth) when approaching the coast.

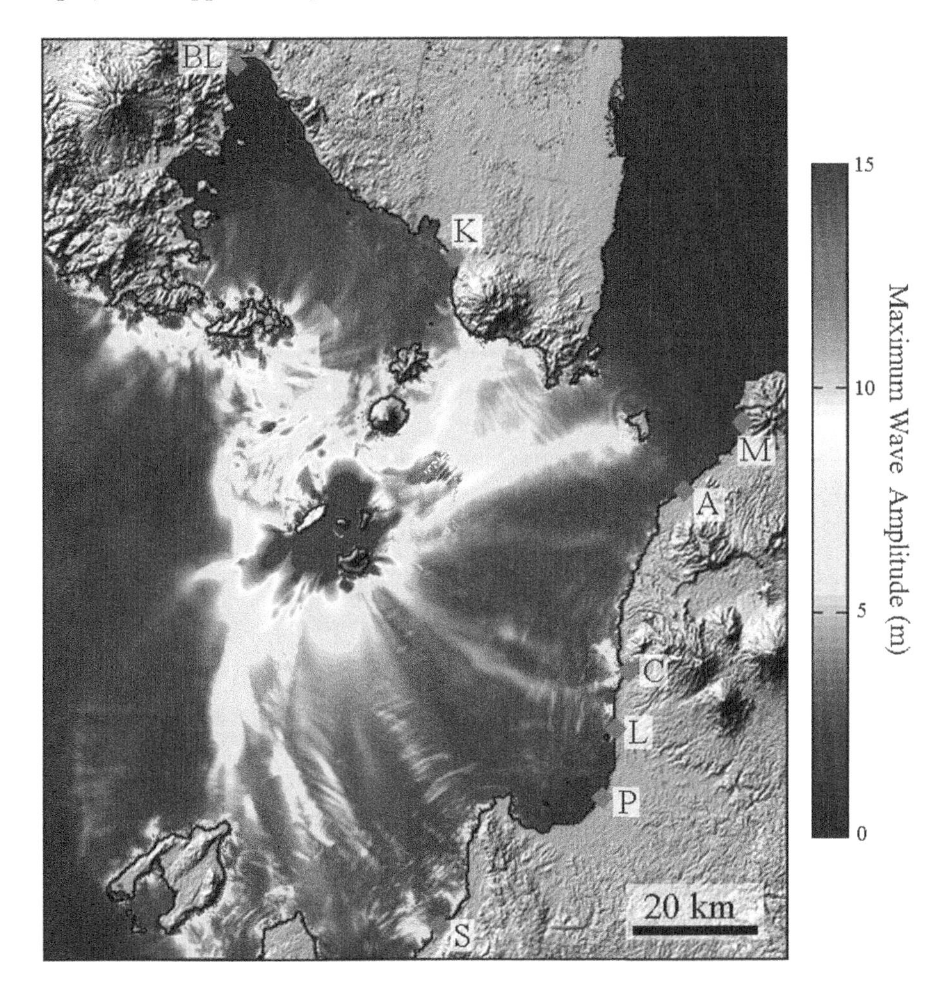

Figure 2.33. *Numerical modeling of a tsunami generated by a landslide of the Anak Krakatau volcano, Sunda Strait, Indonesia (Giachetti et al. 2012). This scenario anticipated as early as 2012 finally occurred on December 22, 2018, with simulated wave heights very close to those observed. Red diamonds and initials show the location of the main coastal cities or important infrastructures around the Sunda Strait. For a color version of this figure, see www.iste.co.uk/lenat/hazards.zip*

In cases where the hydrostatic equilibrium is not conserved, the Boussinesq equations (which are an approximation of the Navier-Stokes equations) must be used to integrate the dispersion phenomena related to complex sources and water depth variations.

At present, the main issues are based on:

– a better integration of tsunami hazard in volcano monitoring systems;

– the improvement of numerical models, especially for complex sources such as pyroclastic flows;

– the development of probabilistic methods of evaluation of the tsunami hazard in volcanic context;

– the sensitization of the populations toward a hazard that is too often considered as marginal.

Volcanic tsunamis are certainly infrequent and their impact is often very localized, but we must be prepared for all scenarios, as the December 22, 2018 Anak Krakatau tsunami and January 15, 2022 Hunga Ha'apai Tonga tsunami unfortunately reminded us.

2.4. References

Allard, P., Aiuppa, A., Bani, P., Métrich, N., Bertagnini, A., Gauthier, P.J. et al. (2016). Prodigious emission rates and magma degassing budget of major, trace and radioactive volatile species from Ambrym basaltic volcano, Vanuatu island Arc. *Journal of Volcanology and Geothermal Research*, 322, 119–143.

Auker, M.R., Sparks, R.S.J., Siebert, L., Crosweller, H.S., Ewert, J. (2013). A statistical analysis of the global historical volcanic fatalities record. *Journal of Applied Volcanology*, 2(1), 1–24.

Barberi, F., Brondi, F., Carapezza, M.L., Cavarra, L., Murgia, C. (2003). Earthen barriers to control lava flows in the 2001 eruption of Mt. Etna. *Journal of Volcanology and Geothermal Research*, 123(1–2), 231–243.

Begét, J. (2000). Volcanic tsunamis. In *Encyplopedia of Volcanoes*, Sigurdsson, H., Houghton, B., McNutt, S.R., Rymer, H., Stix, J. (eds). Academic Press, Cambridge.

Bernabeu, N., Saramito, P., Smutek, C. (2016). Modelling lava flow advance using a shallow-depth approximation for three-dimensional cooling of viscoplastic flows. *Geological Society Special Publication*, 426(1), 409–423.

Bertolaso, G., De Bernardinis, B., Bosi, V., Cardaci, C., Ciolli, S., Colozza, R. et al. (2009). Civil protection preparedness and response to the 2007 eruptive crisis of Stromboli volcano, Italy. *Journal of Volcanology and Geothermal Research*, 182(3/4), 269–277.

Bonaccorso, A., Calvari, S., Garfì, G., Lodato, L., Patanè, D. (2003). Dynamics of the December 2002 flank failure and tsunami at Stromboli volcano inferred by volcanological and geophysical observations. *Geophysical Research Letters*, 30(18). doi.org/10.1029/2003GL017702.

Bonadonna, C. and Costa, A. (2013). Plume height, volume, and classification of explosive volcanic eruptions based on the Weibull function. *Bulletin of Volcanology*, 75(8), 742.

Bonadonna, C., Genco, R., Gouhier, M., Pistolesi, M., Cioni, R., Alfano, F. et al. (2011). Tephra sedimentation during the 2010 Eyjafjallajkull eruption (Iceland) from deposit, radar, and satellite observations. *Journal of Geophysical Research: Solid Earth*, 116(12), 12202.

Brönnimann, S. and Krämer, D. (2016). Tambora and the "Year Without a Summer" of 1816. A perspective on Earth and Human Systems Science. *Geographica Bernensia*, G90, 48.

Brown, R.J. and Andrews, G.D.M. (2015). Deposits of pyroclastic density currents. In *The Encyclopedia of Volcanoes*, 2nd edition, Sigurdsson, H., Houghton, B., McNutt, S.R., Rymer, H., Stix, J. (eds). Academic Press, Amsterdam.

Brown, R.J., Bonadonna, C., Durant, A.J. (2012). A review of volcanic ash aggregation. *Physics and Chemistry of the Earth*, 45/46, 65–78.

Carey, S. and Bursik, M. (2015). Volcanic plumes. In *The Encyclopedia of Volcanoes*, 2nd edition, Sigurdsson, H., Houghton, B., McNutt, S.R., Rymer, H., Stix, J. (eds). Academic Press, New York.

Carracedo, J.C., Day, S.J., Guillou, H., Pérez Torrado, F.J. (1999). Giant quaternary landslides in the evolution of La Palma and El Hierro, Canary Islands. *Journal of Volcanology and Geothermal Research*, 94(1–4), 169–190.

Cervelli, P., Segall, P., Johnson, K., Lisowskl, M., Miklius, A. (2002). Sudden aseismic fault slip on the south flank of Kilauea volcano. *Nature*, 415(6875), 1014–1018.

Chevrel, M.O., Pinkerton, H., Harris, A.J.L. (2019). Measuring the viscosity of lava in the field: A review. *Earth-Science Reviews*, 196, 102852.

Chevrel, M.O., Favalli, M., Villeneuve, N., Harris, A., Fornaciai, A., Richter, N., Derrien, A., Boissier, P., Di Muro, A., Peltier, A. (2021). Lava flow hazard map of Piton de la Fournaise volcano. *Natural Hazards in Earth System Sciences*, 21, 1–22.

Cole, S.E., Cronin, S.J., Sherburn, S., Manville, V. (2009). Seismic signals of snow-slurry lahars in motion: 25 September 2007, Mt Ruapehu, New Zealand. *Geophysical Research Letters*, 36(9), 9.

Cole, P.D., Neri, A., Baxter, P.J. (2015). Hazards from pyroclastic density currents. In *The Encyclopedia of Volcanoes*, 2nd edition, Sigurdsson, H., Houghton, B., McNutt, S.R., Rymer, H., Stix, J. (eds). Academic Press, Amsterdam.

Cooke, R.J.S. (1981). Eruptive history of the volcano at Ritter Island. *Cooke-Ravian Volume of Volcanological Papers*, 10, 115–123.

Coppola, D., Laiolo, M., Franchi, A., Massimetti, F., Cigolini, C., Lara, L.E. (2017). Measuring effusion rates of obsidian lava flows by means of satellite thermal data. *Journal of Volcanology and Geothermal Research*, 347, 82–90.

Costa, J.E. (1988). Rheologic, geomorphic, and sedimentologic differentiation of water floods, hyperconcentrated flows, and debris flows. In *Flood Geomorphology*, Baker, V.R., Kochel, R.C., Patton, P.C. (eds). John Wiley & Sons, Hoboken.

Costa, A., Smith, V.C., Macedonio, G., Matthews, N.E. (2014). The magnitude and impact of the Youngest Toba Tuff super-eruption. *Frontiers in Earth Science*, 2(16). doi.org/10.3389/feart.2014.00016.

Cuomo, S. (2020). Modelling of flowslides and debris avalanches in natural and engineered slopes: A review. *Geoenvironmental Disasters*, 7(1), 1.

D'Alessandro, W. (2006). Human fluorosis related to volcanic activity: A review. In *WIT Transactions on Biomedicine and Health*, Volume 10, Kungolos, A.G., Brebbia, C.A., Samaras C.P. (eds). WIT Press, Ashurst.

De Bélizal, E. (2012). Les corridors de lahars, des espaces entre risque et 146+ ressource. PhD Thesis, Université Paris 1 Pantheon-Sorbonne, Paris.

Del Negro, C., Fortuna, L., Vicari, A. (2005). Modelling lava flows by Cellular Nonlinear Networks (CNN): Preliminary results. *Nonlinear Processes in Geophysics*, 12(4), 505–513.

Del Negro, C., Cappello, A., Neri, M., Bilotta, G., Hérault, A., Ganci, G. (2013). Lava flow hazards at Mount Etna: Constraints imposed by eruptive history and numerical simulations. *Scientific Reports*, 3(3493), 1–8.

Delmelle, P., Stix, J., Baxter, P., Garcia-Alvarez, J., Barquero, J. (2002). Atmospheric dispersion, environmental effects and potential health hazard associated with the low-altitude gas plume of Masaya volcano, Nicaragua. *Bulletin of Volcanology*, 64(6), 423–434.

Derrien, A. (2019). Apports des techniques photogrammétriques à l'étude du dynamisme des structures volcaniques du Piton de la Fournaise. PhD Thesis, Université de Paris, Paris.

Dessert, C., Dupré, B., François, L.M., Schott, J., Gaillardet, J., Chakrapani, G., Bajpai, S. (2001). Erosion of Deccan Traps determined by river geochemistry: Impact on the global climate and the 87Sr/86Sr ratio of seawater. *Earth and Planetary Science Letters*, 188(3–4), 459–474.

Dietterich, H.R., Poland, M.P., Schmidt, D.A., Cashman, K.V., Sherrod, D.R., Espinosa, A.T. (2012). Tracking lava flow emplacement on the east rift zone of Kilauea, Hawaii, with synthetic aperture radar coherence. *Geochemistry, Geophysics, Geosystems*, 13(5), 5001.

Doyle, E.E., Cronin, S.J., Cole, S.E., Thouret, J.C. (2010). The coalescence and organization of lahars at Semeru volcano, Indonesia. *Bulletin of Volcanology*, 72(8), 961–970.

Doyle, E.E., Cronin, S.J., Thouret, J.C. (2011). Defining conditions for bulking and debulking in lahars. *Bulletin of the Geological Society of America*, 123(7–8), 1234–1246.

Dragoni, M., Bonafede, M., Boschi, E. (1986). Downslope flow models of a Bingham liquid: Implications for lava flows. *Journal of Volcanology and Geothermal Research*, 30(3/4), 305–325.

Druitt, T.H., Calder, E.S., Cole, P.D., Hoblitt, R.P., Loughlin, S.C., Norton, G.E. et al. (2002). Small-volume, highly mobile pyroclastic flows formed by rapid sedimentation from pyroclastic surges at Soufrière Hills Volcano, Montserrat: An important volcanic hazard. In *Geological Society Memoir*, 21, Druitt, T.H., Kokelaar, B.P. (eds). Geological Society, London.

Dufek, J., Ongaro, T.E., Roche, O. (2015). Pyroclastic density currents: Processes and models. In *Encyplopedia of Volcanoes*, 2nd edition, Sigurdsson, H., Houghton, B., Mc Nutt, S.R., Rymer, H., Stix, J. (eds). Academic Press, Amsterdam.

Engwell, S. and Eychenne, J. (2016). Contribution of fine ash to the atmosphere from plumes associated with pyroclastic density currents. In *Volcanic Ash: Hazard Observation*, Cashman, K., Ricketts, H., Rust, A., Watson, M. (eds). University of Bristol, Bristol.

Ewing, M. and Press, F. (1955). Tide-gage disturbances from the great eruption of Krakatoa. *Eos, Transactions American Geophysical Union*, 36(1), 53–60.

Favalli, M., Pareschi, M.T., Neri, A., Isola, I. (2005). Forecasting lava flow paths by a stochastic approach. *Geophysical Research Letters*, 32(3), 1–4.

Froger, J.-L., Famin, V., Cayol, V., Augier, A., Michon, L., Lénat, J.-F. (2015). Time-dependent displacements during and after the April 2007 eruption of Piton de la Fournaise, revealed by interferometric data. *Journal of Volcanology and Geothermal Research*, 296. doi.org/10.1016/j.jvolgeores.2015.02.014.

Gauthier, P.J., Sigmarsson, O., Gouhier, M., Haddadi, B., Moune, S. (2016). Elevated gas flux and trace metal degassing from the 2014–2015 fissure eruption at the Bárarbunga volcanic system, Iceland. *Journal of Geophysical Research: Solid Earth*, 121(3), 1610–1630.

Giachetti, T., Paris, R., Kelfoun, K., Ontowirjo, B. (2012). Tsunami hazard related to a flank collapse of Anak Krakatau Volcano, Sunda Strait, Indonesia. *Geological Society Special Publication*, 361(1), 79–90.

Glicken, H. (1996). Rockslide-debris avalanche of May 18, 1980, Mount St. Helens volcano, Washington. Open-file Report 96-677, U.S. Geological Survey, Reston.

Gouhier, M. and Paris, R. (2019). SO2 and tephra emissions during the December 22, 2018 Anak Krakatau flank-collapse eruption. *Volcanica*, 2(2), 91–103 [Online]. Available at: https://doi.org/10.30909/vol.02.02.91103.

Gouhier, M., Guéhenneux, Y., Labazuy, P., Cacault, P., Decriem, J., Rivet, S. (2016). HOTVOLC: A web-based monitoring system for volcanic hot spots. *Geological Society London, Special Publications*, 426(1), 223–241.

Gouhier, M., Eychenne, J., Azzaoui, N., Guillin, A., Deslandes, M., Poret, M. et al. (2019). Low efficiency of large volcanic eruptions in transporting very fine ash into the atmosphere. *Scientific Reports*, 9(1), 1449.

Gudmundsson, M.T., Thordarson, T., Hoskuldsson, A., Larsen, G., Bjornsson, H., Prata, F.J. et al. (2012). Ash generation and distribution from the April–May 2010 eruption of Eyjafjallajökull, Iceland. *Scientific Reports*, 2, 572.

Guffanti, M., Casadevall, T.J., Budding, K. (2011). Encounters of aircraft with volcanic ash clouds: A compilation of known incidents, 1953–2009. The threat of volcanic ash to aviation [Online]. Available at: https://pubs.usgs.gov/ds/545/DS545.pdf.

Hall, M.L., Robin, C., Beate, B., Mothes, P., Monzier, M. (1999). Tungurahua Volcano, Ecuador: Structure, eruptive history and hazards. *Journal of Volcanology and Geothermal Research*, 91(1–2), 1–21.

Harris, A.J.L. (2015). Basaltic lava flow hazard. In *Volcanic Hazards, Risks and Disasters*, Shroder, J.F., Papale, P. (eds). Elsevier, Amsterdam. doi.org/10.1016/B978-0-12-396453-3.00002-2.

Harris, A.J.L. and Rowland, S.K. (2001). FLOWGO: A kinematic thermo-rheological model for lava flowing in a channel. *Bulletin of Volcanology*, 63(1), 20–44.

Harris, A.J.L., De Groeve, T., Garel, F., Carn, S.A. (2016). *Detecting, Modelling and Responding to Effusive Eruptions*. Geological Society, London.

Harris, A.J.L., Chevrel, M.O., Coppola, D., Ramsey, M.S., Hrysiewicz, A., Thivet, S. et al. (2019). Validation of an integrated satellite-data-driven response to an effusive crisis: The April–May 2018 eruption of Piton de la Fournaise. *Annals of Geophysics*, 62(2). doi.org/10.4401/ag-7972.

Holasek, R.E. and Self, S. (1995). GOES weather satellite observations and measurements of the May 18, 1980, Mount St. Helens eruption. *Journal of Geophysical Research*, 100(B5), 8469–8487.

Horwell, C.J. and Baxter, P.J. (2006). The respiratory health hazards of volcanic ash: A review for volcanic risk mitigation. *Bulletin of Volcanology*, 69(1), 1–24.

Hrysiewicz, A. (2019). Caractérisation des déplacements liés aux coulées de lave au Piton de la Fournaise à partir de données InSAR. PhD Thesis, Université Clermont Auvergne, Clermont-Ferrand.

Hungr, O. and Jakob, M. (2005). *Debris Flows Hazards and Related Phenomena*. Praxis/Springer, New York.

Itakura, Y., Inaba, H., Sawada, T. (2005). A debris-flow monitoring devices and methods bibliography. *Natural Hazards and Earth System Science*, 5(6), 971–977.

Iverson, R.M. (2014). Debris flows: Behaviour and hazard assessment. *Geology Today*, 30(1), 15–20.

Jenkins, S.F., Phillips, J.C., Price, R., Feloy, K., Baxter, P.J., Hadmoko, D.S., de Bélizal, E. (2015). Developing building-damage scales for lahars: Application to Merapi volcano, Indonesia. *Bulletin of Volcanology*, 77(9), 1–17.

Keating, B.H. and McGuire, W.J. (2000). Island edifice failures and associated tsunami hazards. *Pure and Applied Geophysics*, 157(6–8), 899–955.

Kelfoun, K. and Druitt, T.H. (2005). Numerical modeling of the emplacement of Socompa rock avalanche, Chile. *Journal of Geophysical Research: Solid Earth*, 110(12), 1–13.

Kelfoun, K. and Vallejo Vargas, S. (2016). VolcFlow capabilities and potential development for the simulation of lava flows. *Geological Society Special Publication*, 426(1), 337–343 [Online]. Available at: https://pubs.geoscienceworld.org/books/book/2014/chapter/16312177/VolcFlow-capabilities-and-potential-development.

Kelfoun, K., Samaniego, P., Palacios, P., Barba, D. (2009). Testing the suitability of frictional behaviour for pyroclastic flow simulation by comparison with a well-constrained eruption at Tungurahua volcano (Ecuador). *Bulletin of Volcanology*, 71(9), 1057–1075.

Lavigne, F. and Thouret, J.-C. (2000). Les lahars : dépôts, origines et dynamique. *Bulletin de La Société géologique de France*, 5(5), 545–557.

Lavigne, F., Thouret, J.-C., Voight, B., Young, K., Lahusen, R., Marso, J. et al. (2000). Instrumental lahar monitoring at Merapi Volcano, Central Java, Indonesia. *Journal of Volcanology and Geothermal Research*, 100(1–4), 457–478.

Legros, F. (2002). The mobility of long-runout landslides. *Engineering Geology*, 63(3–4), 301–331.

Maeno, F. and Imamura, F. (2011). Tsunami generation by a rapid entrance of pyroclastic flow into the sea during the 1883 Krakatau eruption, Indonesia. *Journal of Geophysical Research: Solid Earth*, 116(9), 9205.

Manville, V., Major, J.J., Fagents, S.A. (2009). Modeling lahar behavior and hazards. In *Modeling Volcanic Processes: The Physics and Mathematics of Volcanism*, Fagents, S.A., Gregg, T.K.P., Lopes, R.M.C. (eds). Cambridge University Press, Cambridge.

Martin, E. (2018). Volcanic plume impact on the atmosphere and climate: O- and S-isotope insight into sulfate aerosol formation. *Geosciences (Switzerland)*, 8(6), 198–216.

Mastrodicasa, A., Cuenoud, A., Pasquier, M., Carron, P.-N. (2018). Intoxication aiguë au dioxyde de carbone. *Annales françaises de médecine d'urgence*, 8(5), 326–331.

Mead, S.R., Magill, C., Lemiale, V., Thouret, J.-C., Prakash, M. (2016). Quantifying lahar damage using numerical modelling. *Natural Hazards and Earth System Sciences Discussions*, 17, 1–28.

Michiue, M., Hinokidani, O., Miyamoto, K. (1999). Study on the Mayuyama tsunami disaster in 1792. *Proceedings of the 28th IAHR World Congress*, CD-R d141.pdf.

Miller, C.F. and Wark, D.A. (2008). Supervolcanoes and their explosive supereruptions. *Elements*, 4(1), 11–15.

Miyamoto, H. and Papp, K.R. (2004). Rheology and topography control the path of a lava flow: Insight from numerical simulations over a preexisting topography. *Geophysical Research Letters*, 31(16), 2–5.

Moore, J.G., Clague, D.A., Holcomb, R.T., Lipman, P.W., Normark, W.R., Torresan, M.E. (1989). Prodigious submarine landslides on the Hawaiian Ridge. *Journal of Geophysical Research*, 94(B12), 465–484.

Mossoux, S., Kervyn, M., Canters, F. (2019). Assessing the impact of road segment obstruction on accessibility of critical services in case of a hazard. *Natural Hazards and Earth System Sciences*, 19(6), 1251–1263.

Neal, C.A., Brantley, S.R., Antolik, L., Babb, J., Burgess, M. et al. (2018). The 2018 rift eruption and summit collapse of Kīlauea Volcano. *Science*. doi.org/10.1126/science.aav7046.

Neri, A., Bevilacqua, A., Esposti Ongaro, T., Isaia, R., Aspinall, W.P., Bisson, M. et al. (2015). Quantifying volcanic hazard at Campi Flegrei caldera (Italy) with uncertainty assessment 2: Pyroclastic density current invasion maps. *Journal of Geophysical Research: Solid Earth*, 120(4), 2330–2349.

Oehler, J.F., Labazuy, P., Lénat, J.-F. (2004). Recurrence of major flank landslides during the last 2-Ma-history of Reunion Island. *Bulletin of Volcanology*, 66(7), 585–598.

Ongaro, T.E., Clarke, A.B., Voight, B., Neri, A., Widiwijayanti, C. (2012). Multiphase flow dynamics of pyroclastic density currents during the May 18, 1980 lateral blast of Mount St. Helens. *Journal of Geophysical Research: Solid Earth*, 117(B6). doi.org/10.1029/2011JB009081.

Oppenheimer, C., Fischer, T.P., Scaillet, B. (2014). Volcanic degassing: Process and impact. In *Treatise on Geochemistry*, 2nd edition, Holland, H.D., Turekian, K.K. (eds). Elsevier, Oxford.

Orsi, G., Di Vito, M.A., Isaia, R. (2004). Volcanic hazard assessment at the restless Campi Flegrei caldera. *Bulletin of Volcanology*, 66, 514–530.

Papale, P. and Polacci, M. (1999). Role of carbon dioxide in the dynamics of magma ascent in explosive eruptions. *Bulletin of Volcanology*, 60(8), 583–594.

Paris, R. (2015). Source mechanisms of volcanic tsunamis. *Philosophical Transactions of the Royal Society A: Mathematical, Physical and Engineering Sciences*, 373(2053). doi.org/10.1098/rsta.2014.0380.

Paris, R., Switzer, A.D., Belousova, M., Belousov, A., Ontowirjo, B., Whelley, P.L., Ulvrova, M. (2014a). Volcanic tsunami: A review of source mechanisms, past events and hazards in Southeast Asia (Indonesia, Philippines, Papua New Guinea). *Natural Hazards*, 70(1), 447–470.

Paris, R., Wassmer, P., Lavigne, F., Belousov, A., Belousova, M., Iskandarsyah, Y. et al. (2014b). Coupling eruption and tsunami records: The Krakatau 1883 case study, Indonesia. *Bulletin of Volcanology*, 76(4), 1–23.

Paris, R., Ramalho, R.S., Madeira, J., Ávila, S., May, S.M., Rixhon, G. et al. (2018). Mega-tsunami conglomerates and flank collapses of ocean island volcanoes. *Marine Geology*, 395, 168–187.

Pastor, M., Haddad, B., Sorbino, G., Cuomo, S., Drempetic, V. (2009). A depth-integrated, coupled SPH model for flow-like landslides and related phenomena. *International Journal for Numerical and Analytical Methods in Geomechanics*, 33(2), 143–172.

Pelinovsky, E., Choi, B.H., Stromkov, A., Didenkulova, I., Kim, H.S. (2005). Analysis of tide-gauge records of the 1883 Krakatau tsunami. In *Tsunamis: Case Studies and Recent Developments*, Satake, K. (ed.), Springer, The Netherlands.

Pierson, T.C., Janda, R.J., Thouret, J.C., Borrero, C.A. (1990). Perturbation and melting of snow and ice by the 13 November 1985 eruption of Nevado del Ruiz, Colombia, and consequent mobilization, flow and deposition of lahars. *Journal of Volcanology and Geothermal Research*, 41(1–4), 17–66.

Pierson, T.C., Wood, N.J., Driedger, C.L. (2014). Reducing risk from lahar hazards: Concepts, case studies, and roles for scientists. *Journal of Applied Volcanology*, 3(1). doi.org/10.1186/s13617-014-0016-4.

Pitman, E.B. and Long, L.E. (2005). A two-fluid model for avalanche and debris flows. *Philosophical Transactions of the Royal Society A: Mathematical, Physical and Engineering Sciences*, 363(1832), 1573–1601.

Robock, A. (2000). Volcanic eruptions and climate. *Reviews of Geophysics*, 38, 191–219.

Roche, O., Buesch, D.C., Valentine, G.A. (2016). Slow-moving and far-travelled dense pyroclastic flows during the Peach Spring super-eruption. *Nature Communications*, 7. doi.org/10.1038/ncomms10890.

Rowland, S.K., Garbeil, H., Harris, A.J.L. (2005). Lengths and hazards from channel-fed lava flows on Mauna Loa, Hawai'i, determined from thermal and downslope modeling with FLOWGO. *Bulletin of Volcanology*, 67(7), 634–647.

Sandri, L., Tierz, P., Costa, A., Marzocchi, W. (2018). Probabilistic hazard from pyroclastic density currents in the Neapolitan Area (Southern Italy). *Journal of Geophysical Research: Solid Earth*, 123(5), 3474–3500.

Sarna-Wojcicki, A.M., Shipley, S., Waitt, R.B.J., Dzurisin, D., Wood, S.H. (1981). Areal distribution, thickness, mass, volume, and grain size of air-fall ash from six major eruptions of 1980. *The 1980 Eruptions of Mount St. Helens, Washington*, 1250, 577–600.

Satake, K. and Kato, Y. (2001). The 1741 Oshima-Oshima eruption: Extent and volume of submarine debris avalanche. *Geophysical Research Letters*, 28(3), 427–430.

Schilling, S.P. (2014). Laharz_py: GIS tools for automated mapping of lahar inundation hazard zones. US Geological Survey, Open-File Report Vol. 2014 (2014–1073). doi:10.3133/ofr20141073.

Scollo, S., Bonadonna, C., Manzella, I. (2017). Settling-driven gravitational instabilities associated with volcanic clouds: New insights from experimental investigations. *Bulletin of Volcanology*, 79(6), 39.

Scott, K.M. (1988). Origins, behavior, and sedimentology of lahars and lahar-runout flows in the Toutle-Cowlitz River system: Lahars and lahar-runout flows in the Toutle-Cowlitz River system, Mount St. Helens, Washington. *US Geological Survey Professional Paper*, 1447(A), 1–76.

Scott, K.M., Vallance, J.W., Kerle, N., Macías, J.L., Strauch, W., Devoli, G. (2005). Catastrophic precipitation-triggered lahar at Casita volcano, Nicaragua: Occurrence, bulking and transformation. *Earth Surface Processes and Landforms*, 30(1), 59–79.

Self, S. (2006). The effects and consequences of very large explosive volcanic eruptions. *Philosophical Transactions of the Royal Society A: Mathematical, Physical and Engineering Sciences*, 364(1845), 2073–2097.

Self, S. and Rampino, M.R. (1981). The 1883 eruption of Krakatau, *Nature*, 294, 699–704.

Shea, T. (2017). Bubble nucleation in magmas: A dominantly heterogeneous process? *Journal of Volcanology and Geothermal Research*, 343, 155–170.

Shea, T. and van Wyk de Vries, B. (2008). Structural analysis and analogue modeling of the kinematics and dynamics of rockslide avalanches. *Geosphere*, 4(4), 657–686.

Siebert, L. (1984). Large volcanic debris avalanches: Characteristics of source areas, deposits, and associated eruptions. *J. Volcanol. Geotherm. Res.*, 22(3/4), 163–197.

Siebert, L., Glicken, H., Ui, T. (1987). Volcanic hazards from Bezymianny- and Bandai-type eruptions. *Bulletin of Volcanology*, 49(1), 435–459.

Siebert, L., Simkin, T., Kimberly, P. (2010). *Volcanoes of the World*, 3rd edition. University of California Press, Berkeley.

Simkin, T. and Fiske, R.S. (1983). *Krakatau 1883: The Volcanic Eruption and Its Effects*. Smithsonian Institution Press, Washington, DC.

Solikhin, A., Thouret, J.-C., Gupta, A., Sayudi, D.S., Oehler, J.-F., Liew, S.C. (2015). Effects and behavior of pyroclastic and lahar deposits of the 2010 Merapi eruption based on high-resolution optical imagery. *Procedia Earth and Planetary Science*, 12, 1–10.

Sparks, R.S.J. (1978). The dynamics of bubble formation and growth in magmas: A review and analysis. *Journal of Volcanology and Geothermal Research*, 3(1), 1–37.

Strachey, R. (1888). On the air waves and sounds caused by the eruption of Krakatoa in August 1883. In *Krakatau 1883*, Symons, G. (ed.). Trubner, London.

Streck, M.J. and Grunder, A.L. (1995). Crystallization and welding variations in a widespread ignimbrite sheet; the Rattlesnake Tuff, eastern Oregon, USA. *Bulletin of Volcanology*, 57(3), 151–169.

Symonds, R.B., Rose, W.I., Bluth, G.J.S., Gerlach, T.M. (1994). Volcanic-gas studies: Methods, results, and applications. In *Reviews in Mineralogy and Geochemistry*, Volume 30, Carroll, M.R., Holloway, J.R. (eds). Mineralogical Society of America, Chantilly.

Tang, Y., Tong, D.Q., Yang, K., Lee, P., Baker, B., Crawford, A. et al. (2020). Air quality impacts of the 2018 Mt. Kilauea Volcano eruption in Hawaii: A regional chemical transport model study with satellite-constrained emissions. *Atmospheric Environment*, 237, 117648.

Tedesco, D., Vaselli, O., Papale, P., Carn, S.A., Voltaggio, M., Sawyer, G.M. et al. (2007). January 2002 volcano-tectonic eruption of Nyiragongo volcano, Democratic Republic of Congo. *Journal of Geophysical Research: Solid Earth*, 112(9), 9202.

Thordarson, T. and Self, S. (2003). Atmospheric and environmental effects of the 1783–1784 Laki eruption: A review and reassessment. *Journal of Geophysical Research: Atmospheres*, 108(1), 2042.

Thouret, J.C., Lavigne, F., Kelfoun, K., Bronto, S. (2000). Toward a revised hazard assessment at Merapi volcano, Central Java. *Journal of Volcanology and Geothermal Research*, 100(1–4), 479–502.

Thouret, J.C., Antoine, S., Magill, C., Ollier, C. (2020). Lahars and debris flows: Characteristics and impacts. *Earth-Science Reviews*, 201, 479–502.

Turnbull, B., Bowman, E.T., McElwaine, J.N. (2015). Debris flows: Experiments and modelling. *Comptes rendus physique*, 16(1), 86–96.

Vallance, J.W. and Iverson, R.M. (2015). Lahars and their deposits. In *The Encyclopedia of Volcanoes*, Sigurdsson, H. (ed.). Elsevier/Academic Press, Cambridge.

Verbeek, R.M. (1886). *Krakatau*. Imprimerie de l'État, Jakarta.

Voight, B., Glicken, H., Janda, R.J., Douglass, M. (1981). Catastrophic rockslide avalanche of May 18 (Mount St. Helens). *U.S. Geological Survey Professional Paper*, 1250, 347–377.

Wadge, G., Young, P.A.V., McKendrick, I.J. (1994). Mapping lava flow hazards using computer simulation. *Journal of Geophysical Research*, 99(B1), 489–504.

Wallace, P.J. (2005). Volatiles in subduction zone magmas: Concentrations and fluxes based on melt inclusion and volcanic gas data. *Journal of Volcanology and Geothermal Research*, 140(1–3), 217–240.

Walter, T.R., Haghshenas Haghighi, M., Schneider, F.M., Coppola, D., Motagh, M., Saul, J. et al. (2019). Complex hazard cascade culminating in the Anak Krakatau sector collapse. *Nature Communications*, 10(1), 4339.

Wheeler, D. and Demarée, G. (2005). The weather of the Waterloo campaign 16 to 18 June 1815: Did it change the course of history? *Weather*, 60(6), 159–164.

White, R.V. and Saunders, A.D. (2005). Volcanism, impact and mass extinctions: Incredible or credible coincidences? *Lithos*, 79(3–4), 299–316.

Wilson, T.M. and Kaye, G.D. (2007). Agricultural fragility estimates for volcanic ash fall hazards. *GNS Science Report*, Lower Hutt, New Zealand, 2007(37), 51.

Wilson, L., Sparks, R.S.J., Walker, G.P.L. (1980). Explosive volcanic eruptions IV – The control of magma properties and conduit geometry on eruption column behaviour. *Geophysical Journal of the Royal Astronomical Society*, 63(1), 117–148.

Wilson, C.J.N., Houghton, B.F., McWilliams, M.O., Lanphere, M.A., Weaver, S.D., Briggs, R.M. (1995). Volcanic and structural evolution of Taupo Volcanic Zone, New Zealand: A review. *Journal of Volcanology and Geothermal Research*, 68(1), 1–28.

Wilson, T.M., Stewart, C., Sword-Daniels, V., Leonard, G.S., Johnston, D.M., Cole, J.W. et al. (2012). Volcanic ash impacts on critical infrastructure. *Physics and Chemistry of the Earth*, 45/46, 5–23.

Wilson, G., Wilson, T.M., Deligne, N.I., Cole, J.W. (2014). Volcanic hazard impacts to critical infrastructure: A review. *Journal of Volcanology and Geothermal Research*, 286, 148–182.

Wilson, T.M., Jenkins, S.F., Stewart, C. (2015). Volcanic ash fall impacts. In *Volcanic Hazards and Risk Global*, Loughlin, S.C., Sparks, R.S.J., Brown, S.K., Jenkins, S.F., Vye-Brown, C. (eds). Cambridge University Press, Cambridge.

Wilson, G., Wilson, T.M., Deligne, N.I., Blake, D.M., Cole, J.W. (2017). Framework for developing volcanic fragility and vulnerability functions for critical infrastructure. *Journal of Applied Volcanology*, 6(1), 1–24.

Winner, W.E. and Mooney, H.A. (1980). Responses of Hawaiian plants to volcanic sulfur dioxide: Stomatal behavior and foliar injury. *Science*, 210(4471), 789–791.

Wright, J.V., Smith, A.L., Roobol, M.J., Mattioli, G.S., Fryxell, J.E. (2016). Distal ash hurricane (pyroclastic density current) deposits from a ca. 2000 yr B.P. Plinian-style eruption of Mount Pelée, Martinique: Distribution, grain-size characteristics, and implications for future hazard. *Geological Society of America Bulletin*, 128(5/6), 777–791.

van Wyk de Vries, B. and Delcamp, A. (2015). Volcanic debris avalanches. *Landslide Hazards, Risks, and Disasters*, 131–157. doi.org/10.1016/B978-0-12-396452-6.00005-7.

Yavari-Ramshe, S. and Ataie-Ashtiani, B. (2016). Numerical modeling of subaerial and submarine landslide-generated tsunami waves – Recent advances and future challenges. *Landslides*, 13(6), 1325–1368.

Yokoyama, I. (1987). A scenario of the 1883 Krakatau tsunami. *Journal of Volcanology and Geothermal Research*, 34(1–2), 123–132.

Zambri, B., Robock, A., Mills, M.J., Schmidt, A. (2019). Modeling the 1783–1784 Laki eruption in Iceland 2: Climate impacts. *Journal of Geophysical Research: Atmospheres*, 124(13), 6770–6790.

Zanchetta, G., Sulpizio, R., Pareschi, M.T., Leoni, F.M., Santacroce, R. (2004). Characteristics of May 5–6, 1998 volcaniclastic debris flows in the Sarno area (Campania, southern Italy): Relationships to structural damage and hazard zonation. *Journal of Volcanology and Geothermal Research*, 133(1–4), 377–393.

3

Assessment, Delineation of Hazard Zones and Modeling of Volcanic Hazards

Jean-Claude THOURET[1] and Sylvain CHARBONNIER[2]
[1] *Laboratoire Magmas et Volcans, CNRS, IRD, OPGC, Université Clermont Auvergne, Clermont-Ferrand, France*
[2] *School of Geosciences, University of South Florida, Tampa, United States*

3.1. Introduction

Despite technological and scientific advances in volcanology, recent eruptions (Agung in 2017–2018; Puna in Hawaii in May–July 2018; and Fuego in June 2018) remind us that volcanic events are among the most frequent and destructive natural hazards. Since 1500 CE, eruptions have caused an estimated 278,500 casualties (Auker et al. 2013; Brown et al. 2017), far fewer, however, than other natural disasters such as earthquakes, atmospheric events or epidemics. Nevertheless, the destructive potential of eruptions on property is considerable, while the induced changes to the landscape and environment are notorious. Eruptions can have a high economic impact in territories far from an active volcano, even when their magnitudes remain modest (Volcanic Explosivity Index, VEI < 3)[1]; consider the ash of Eyjafjallajökhul (Iceland) in continental Europe in April 2010. The modest eruption of Nevado del Ruiz in 1985 in Colombia resulted in an economic loss of 20% of the country's GDP. Over the past 30 years, damage has been rising

1 Find the definition of the index in Chapter 2.

Hazards and Monitoring of Volcanic Activity 1,
coordinated by Jean-François LÉNAT. © ISTE Ltd 2022.

sharply around the world due to the increase in socio-economic vulnerability and the value of assets exposed to natural hazards.

3.2. Terminology

The concepts and content of hazard and risk have evolved over time, especially as several disciplines have employed these concepts by broadening the field of research beyond natural hazards, particularly in the social sciences.

Risk is a concept that combines hazard, exposure, vulnerability and the capacity of society to cope with the potential damage.

Several authors have defined *the hazard* by its occurrence, the area affected, the duration, the magnitude/frequency relationship and a probability.

Exposure includes the location, position and state of each of the elements (inhabitants, buildings, infrastructures, networks and resources) located in the path of the potentially damaging phenomenon.

Exposure is often confused with *vulnerability*. We define the latter as the propensity to suffer damage. Vulnerability is expressed by indicators of vulnerability or fragility and the percentage of probable damage. Vulnerability is a complex set of parameters and factors that are interdependent and that include several domains: structural (type of building), human (people affected, displaced, injured or disabled), functional (damage to the organization of the main economic activities), technical (damage to transport, lifelines and networks), psychological (on the scale of individuals and groups) and circumstantial (time of day or year when the crisis strikes) (D'Ercole and Metzger 2004).

Damage is defined by the value (not only monetary) of the elements exposed, also called *assets*. The latter includes all socio-economic and governance elements.

Prior to the risk analysis, the evaluation of the hazard, the vulnerability and the assets at stake requires several steps of investigation of which the delineation and mapping of areas potentially affected represents one of the main links. The delineation of exposed areas may be useful for stakeholders to take the necessary preventive measures. However, we stress the fact that the intensity and effects of the hazard are not defined once and for all, as they evolve with the training, behavior and resilience of the society exposed.

3.3. Objectives of volcanic hazard assessment and delineation of hazard zones

The assessment and mitigation of volcanic hazards pose a series of challenges, some still insurmountable, others difficult to solve despite progress in understanding the phenomena and in monitoring volcanic activity. Despite the knowledge acquired through systematic observation and modeling of recent volcanic eruptions, hazard mapping and delineation of hazard zones around active volcanoes remains a challenge. Yet, an estimated 800 million people live in hazardous areas within 100 km of active volcanoes (Brown et al. 2017). Key challenges include the following issues:

– An erupting volcano does not present a single hazard; multiple hazards vary in time and space.

– The duration of eruptions, on average from a few weeks to a month, is often long compared to the perception of the inhabitants.

– An active volcanic edifice can present different eruptive styles and generate a variable range and volume of eruptive products.

– Active and dormant volcanoes generate hazards that are prolonged in time and space (e.g. lahars around Pinatubo more than 13 years after the 1991 eruption).

– Knowledge of the eruptive history of active volcanoes (see Chapter 1) is often too sparse to allow a statistical study of the magnitude/ recurrence relationship of events.

This chapter will review three topics: (1) a summary of the major hazards and their effects; (2) methods of delineation of hazard zones around volcanoes and the methodological evolution over the past 30 years; and (3) new modeling approaches and the development of quantitative analyses to improve hazard assessment and risk mitigation.

3.4. The main volcanic hazards and their effects

We consider two classes of volcanic hazards (see Table 3.1):

– direct or associated eruptive hazards;

– post-eruptive hazards.

Chapter 2 of this volume discusses these hazards in more detail.

Type of hazard	Characteristics relevant to hazard assessment	Main effects
	Direct or associated eruptive hazards	
Lava flows	Channel and tube formation Front speed, flow rate, duration Topography	Crushing, fires Long-term burial Toxic gases and vog mist
Lava domes	Growth rate Morphology and topography Can remain unstable for a long time after the end of an eruption	Falling blocks Explosive activity, pyroclastic flows Remobilized non-consolidated products (lahars or debris flows)
Ballistic projections	Little or no influence of wind	Impact (distance <10 km)
Ash plumes and aerosols	Volume, flow, duration, grain-size distribution, composition, density, height	Spillover to very large areas Burial Roof collapses Damage to electrical and mechanical equipment Health, aviation, climate Disruption of the troposphere and stratosphere
Pyroclastic flows (PFs)	High temperature (100°C–400°C) High speed (>tens of m/s) Flow in valleys but can cross topographic obstacles	Burial Destruction of buildings (due to their mass and speed) Combustion, fires Burning and asphyxiation of living beings
Shock waves	Associated with explosions	Damage to buildings. Depending on the pressure: breakage of windows, doors, collapse of walls, etc. Damage to eardrums up to lethal effects

Type of hazard	Characteristics relevant to hazard assessment	Main effects
	Direct or associated eruptive hazards	
Collapses, debris avalanches	Can be large (up to 50 km^3) Very fast (100–300 m/s) Can travel long distances (up to 100 km)	Destruction and burial River damming Tsunamis
Caldera collapse	Difference between the calderas of basaltic volcanoes (associated with large effusive eruptions) and those of more acidic volcanoes (associated with very voluminous explosive phases)	Strong earthquakes Damage associated with concomitant eruptions
Syn-eruptive lahars	Volume and velocity increase with distance Aggradation, incision, channel migration Overflow and avulsion possible Acidity (low pH) inflicting burns	Destructions Distal sedimentation Contamination
Jökhulhlaup	Very high flow and capacity Difficult to forecast	Flooding, destruction, accumulation and erosion
Gases	Emitted into the atmosphere during eruptions, by hydrothermal systems or by permanent and/or diffuse degassing Mainly H_2O, CO_2, SO_2, H_2, CO and, in smaller quantities, H_2S, HCl, HF, He, etc. Dispersed in the atmosphere as gases, acid aerosols, compounds attached to tephra, salt particles	Short distance from sources: – effects related to the toxicity of gases; – acid fog and rain; – pollution of water reservoirs and poisoning of pastures More global effects (chemistry and solar radiation absorption in the stratosphere; greenhouse effect in the troposphere)

Type of hazard	Characteristics relevant to hazard assessment	Main effects
Direct or associated eruptive hazards		
Ground deformation and earthquakes	Magma intrusion Variation of pressure within hydrothermal systems Volcano-tectonic phenomena	Ground deformation and disruption (fractures) Repeated, small to moderate sized earthquakes may produce damage (rockfalls from unstable, steep slopes) High magnitude earthquakes induce effects that are similar to those of tectonic earthquakes
Volcanic tsunamis	Water displacement due to fast currents or volcano-tectonic phenomena	Devastation, flooding, coastal erosion
Secondary explosions	On lava flows and pyroclastic flows once subsurface water is present	Danger is localized Modify local drainage
Post-eruptive hazards		
Post-eruptive lahars	Accumulation of unconsolidated products Topography Rainfall	Same as the syn-eruptive lahars
Debris avalanches and landslides not associated with eruptive activity	Gravitational instability linked to hydrothermal alteration, erosion, rainfall episodes of exceptional amplitude, strong earthquakes	Same as their syn-eruptive counterparts
Volcanic dam rupture	Caldera or valley dammed lakes by volcanic products	Flooding, destruction, accumulation and erosion

Table 3.1. *The main volcanic hazards: characteristics, effects and examples (modified from Blong (2000), Thouret (2004) and Tilling (2005))*

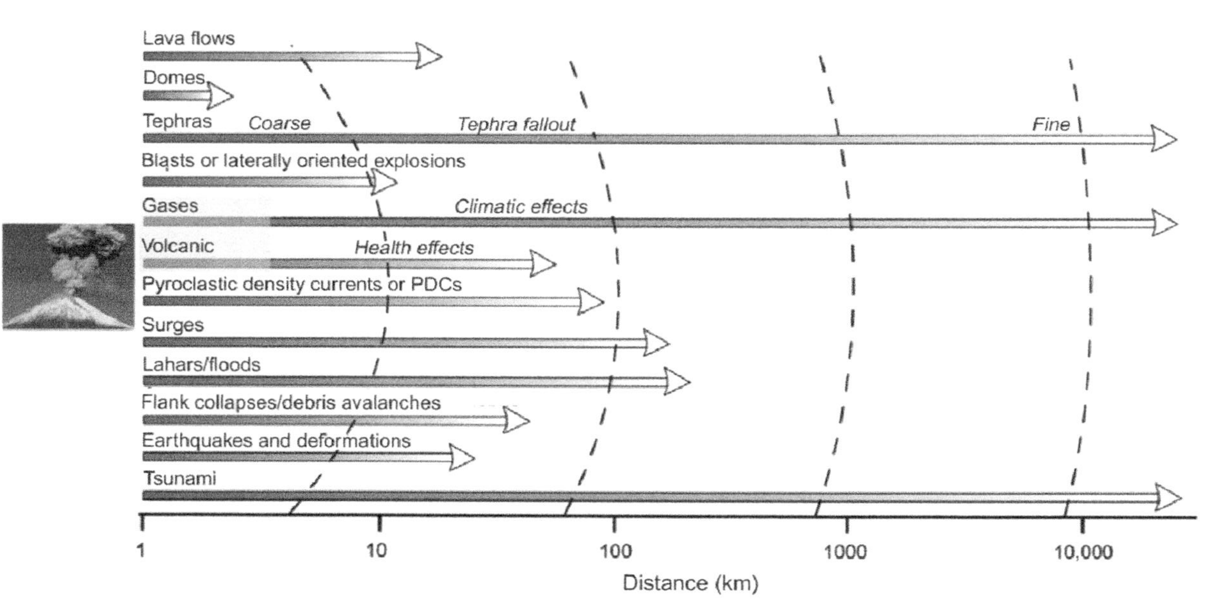

Figure 3.1. *Volcanic hazards and propagation distances for major direct, indirect and associated eruptive events (modified from Chester et al. (2000))*

3.4.1. *Temporal and spatial scales of hazards*

Most damaging eruptive events occur within a radius of 5–10 km of an active volcano or as far as 20 km in the event of large eruptions with a VEI > 3 (see Figure 3.1). Some of these hazardous events, such as lahars and fallout, can reach remarkable distances around vents: 100 km for the former along the drainage network (Pierson et al. 1990), hundreds to thousands of kilometers for fine ash and aerosols. Volcanogenic tsunamis, like those generated by the 1883 eruption of Krakatau, can cause damage to coastlines thousands of kilometers away from the volcano. Even more extensive is the damage indirectly induced by the injection of volcanic gases, mainly SO_2, into the stratosphere with transient changes in climate.

We can distinguish between recurrent hazards, such as tephra (fallout) or lava flows, and rarer ones such as debris avalanches. Post-eruptive hazards, such as those caused by the remobilization of tephra and volcaniclastic sediments, can occur during rainy seasons over one or two decades around volcanoes located in intertropical and temperate latitudes. The hazards become more sporadic when it comes, for example, to the rupture of lakes created by eruptions (crater lakes or valley dams). In total, such volcanic hazards of multiple origins and affecting a large area over long periods of time often have considerable effects. The diversity of their processes, their screening distances and the impacts caused put them almost on a par with certain climatic phenomena.

3.4.2. *Existing hazard classifications and their criteria*

Previously based essentially on the geology of associated deposits, the description of hazards on active volcanoes developed at the beginning of the 1980s by considering the almost synchronous dynamic aspects of several phenomena. The best known example is that observed during the May 18, 1980 eruption of Mount St. Helens, which successively combined: M = 5.1 earthquake, flank collapse, debris avalanche, directed lateral explosion or blast, column and plume of gas and particles, fallout, pyroclastic density currents and primary lahars. Similarly, the hazard assessment recognized the multiplicity of "cascading" effects. For example, the fallout and pyroclastic density currents emplaced from Pinatubo after June 15, 1991 induced lahars.

In addition, there are indirect phenomena, which have not much been considered before, such as the long-distance ash dispersion and their effects on air traffic or on health. Enhanced sedimentation and erosion processes, caused by lahars and other flows remobilizing voluminous unconsolidated pyroclastic deposits, highlighted a new category of mixed and geomorphic hazards in the early 1990s (Pierson and Major 2014). The cases of Pinatubo (Newhall and Punongbayan 1996) and Chaíten (Major and Lara 2013) have thus highlighted the morphological role of extreme sedimentation in the very short term, with tephra accumulation thick enough to inhibit rainfall infiltration capacity, and the rapid response of watersheds subjected to intense runoff. In the case of volcanoes mechanically weakened by hydrothermal alteration or gravitationally unstable, exceptional hydrometeorological phenomena can generate landslides that turn into debris flows downstream. One example is Casita volcano (Nicaragua) during Hurricane Mitch in 1988 (Scott et al. 2005).

Today, hazard classifications integrate all eruptive and post-eruptive phenomena, non-eruptive processes and multiple characteristics, including geomorphological, hydrological and monitoring characteristics (this approach is sometimes referred to as "holistic") (see Box 3.1, "Eruptive scenarios"). Recent classifications take into account the specificity of the volcanic hazards in relation to other natural phenomena, as well as the interactions of eruptive processes with the surrounding environment.

3.5. Multi-hazard delineation methods for volcanoes

Hazard assessment and delineation consists of estimating the degree of exposure of an entire region and then delineating the areas potentially affected or subject to indirect or delayed disturbances from eruptive processes. Mapping and delineation methods have evolved considerably over the last 30 years or so. Initially geological, empirical and deterministic, these methods evolved with the use of quantified field data, numerical simulation codes, remote sensing techniques, monitoring and laboratory experiments.

Three major innovations have improved hazard mapping over the last 20 years:

– the incorporation of numerical or statistical models and simulations;

– the introduction of the concept of scenario, based on the triple relationship between magnitude, frequency and intensity;

– the more recent consideration of the impacts of specific processes on the habitat and infrastructures whose geotechnical characteristics are defined. This is the case for the dynamic pressure exerted by pyroclastic flows (PFs) and lahars, resulting from the inventory of damage recorded after a major eruption such as that of Merapi in 2010 (Jenkins et al. 2013; Thouret et al. 2020).

3.5.1. *Specificity and complexity of volcanic hazard delineation*

Delineating areas likely to be affected by volcanic eruptions is a difficult task compared with other natural hazards. Several characteristics make these hazards complicated to assess and delineate (Stieltjes 2004):

– Volcanic activity is a phenomenon with random frequency whose return period is generally quite long (decades) to very long (millennia), except for volcanoes whose activity is permanent or more frequent. The low frequency of large eruptions implies a relative scarcity of information available for crisis management, as well as the loss of memory of the associated impacts on affected inhabitants.

– Volcanic hazards are manifold and long-lasting, because post-eruptive phenomena can be triggered in the medium and long term (months to decades), including during intervals of relative quiescence between long eruptive episodes, and can occur in unexpected areas.

– Volcanic activity affects both humans and all environments, living, built and environmental. The consequences and damage to vulnerable environments can reach a whole continent, or even the entire planet in the case of fallout and atmospheric effects triggered by "super-eruptions" (Self 2006).

– The prevention measures do not have visible effects in the short term, which slow down the decision-makers to consider both the hazard and exposure into risk planning and mitigation.

– Understanding and monitoring of active volcanoes remain insufficient in developing countries (DCs), where active and dangerous volcanoes are numerous. In these DCs, catastrophic eruptions (e.g. Pinatubo in 1991 and Fuego in 2018) have occurred on volcanoes whose past eruptive record was very limited and monitoring almost non-existent or poorly developed.

3.5.2. *Principles of hazard delineation*

A quantitative method of hazard delineation is based on a protocol for which several input parameters must be defined (Stieltjes 2004): the studied period, the most likely damaging eruption and a quantitative scale of intensity and frequency for each hazard.

To expand the hazard delineation to risk delineation, we can add exposure indices based on matrices of intensity/frequency/damage or victims/recurrence for any type of element or asset at stake. These assets are located, described and classified according to their value by geographical site (valley, interfluve, piedmont, flood plain, etc.), by sector of economic activity (primary, secondary or tertiary) and by group of exposed elements (administrative, school, commercial, cultural, sports, religious, etc.).

We can distinguish several types of hazard zone delineation:

– Classical delineation, based on geological and geomorphological approaches, defines geographical areas exposed to each eruptive hazard in a deterministic way and according to the record of recent eruptive activity.

– The more elaborate delineation ranks areas according to a matrix resulting from the combination of exposure with intensity indices for each phenomenon. These parameters are either identified from the past eruptive record or simulated or derived from a statistical processing of the identified impacts.

– The delineation established based on eruptive scenarios is based on mathematical tools, the results of laboratory experiments, numerical simulations and feedback from recent crises. It is similar to risk delineation if the geographical areas are classified according to the scenarios, using statistical matrices combining three indices: population, exposure and vulnerability of populations, housing, infrastructures, networks (fluids, transport and lifelines) and monetary issues.

3.5.3. *The graphic expression of delineation of hazard zones: the hazard maps*

We express the areas likely to be affected by volcanic activity delineated in the form of hazard-zone maps, now integrated into GIS (Geographic Information System) that facilitate the georeferencing, integration and visualization of numerous information on several layers, their update and their dissemination by current means of communication.

A census of 120 published maps allowed Calder et al. (2015) to distinguish between five types of maps (see Figure 3.2).

– The more common deterministic map relies on geology and indicates the distribution of deposits from past eruptions, and sometimes their recurrence. Such maps either consider a single or several hazards during an eruption.

– The integrated qualitative or global hazard map, with the integration of several hazards in the same area, presents zones of often quasi-concentric intensity, as for the hazard levels that result from the amalgamation of information on the typology, extension and frequency of hazards.

– The scenario map, which is more detailed, is the result of simulations of certain hazards based on some input parameters and using a digital terrain model (DTM).

– The stochastic map is based on probabilistic evaluations of hazards from databases (physical parameters of the phenomena, results of experiments or simulations) and areas. It includes the evaluation of the vulnerability of the elements at stake (buildings, infrastructure and networks) and event-based decision trees.

– The administrative map, based on both hazard information and census data, is oriented toward prevention measures (evacuation routes, shelters and strategic resources in case of crisis). This type of map also designates prohibited areas and temporary or provisional access areas.

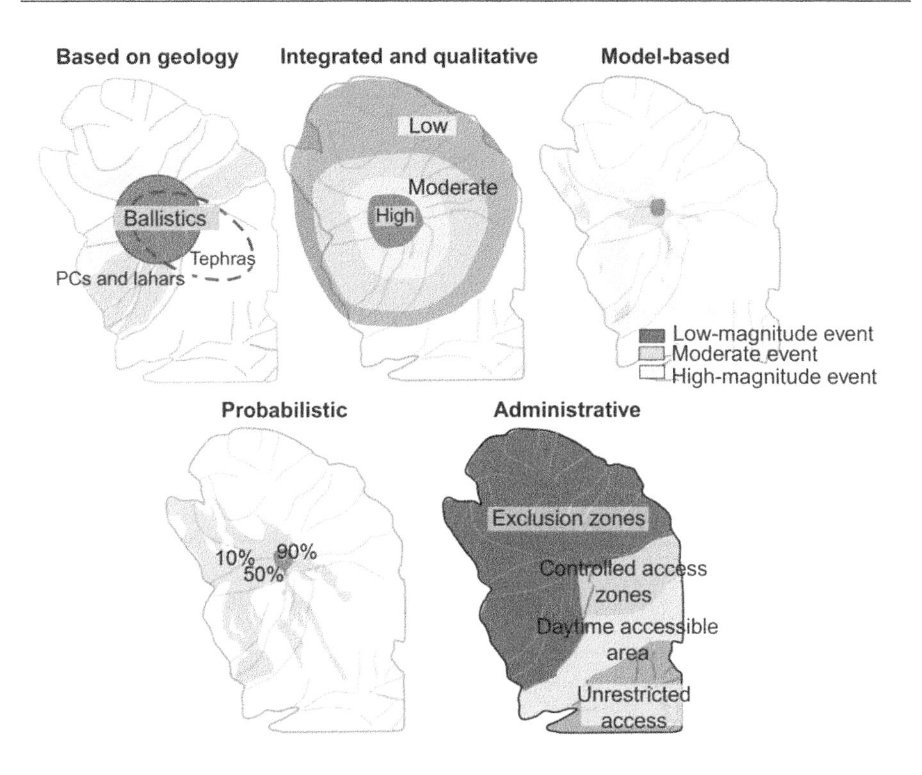

Figure 3.2. *Examples of the five predominant types of hazard maps (modified from Calder et al. (2015)). For a color version of this figure, see www.iste.co.uk/lenat/hazards.zip*

COMMENT ON FIGURE 3.2.– *The application volcano is fictitious but resembles Soufrière Hills on the island of Montserrat. It allows us to illustrate the five types of hazard maps on the same topography. Each map represents a different group of input parameters (explained in the text).*

Other types of maps can be used for specific purposes, such as rapid hazard-zone sketch maps during a crisis, geared toward emergency response. In contrast to traditional hazard-zone maps, which volcanologists build during quiescent periods of activity and show potentially affected areas in the long term, temporary maps, made quickly during a crisis, communicate emergency measures in response to the imminence of expected events (see Figure 3.3).

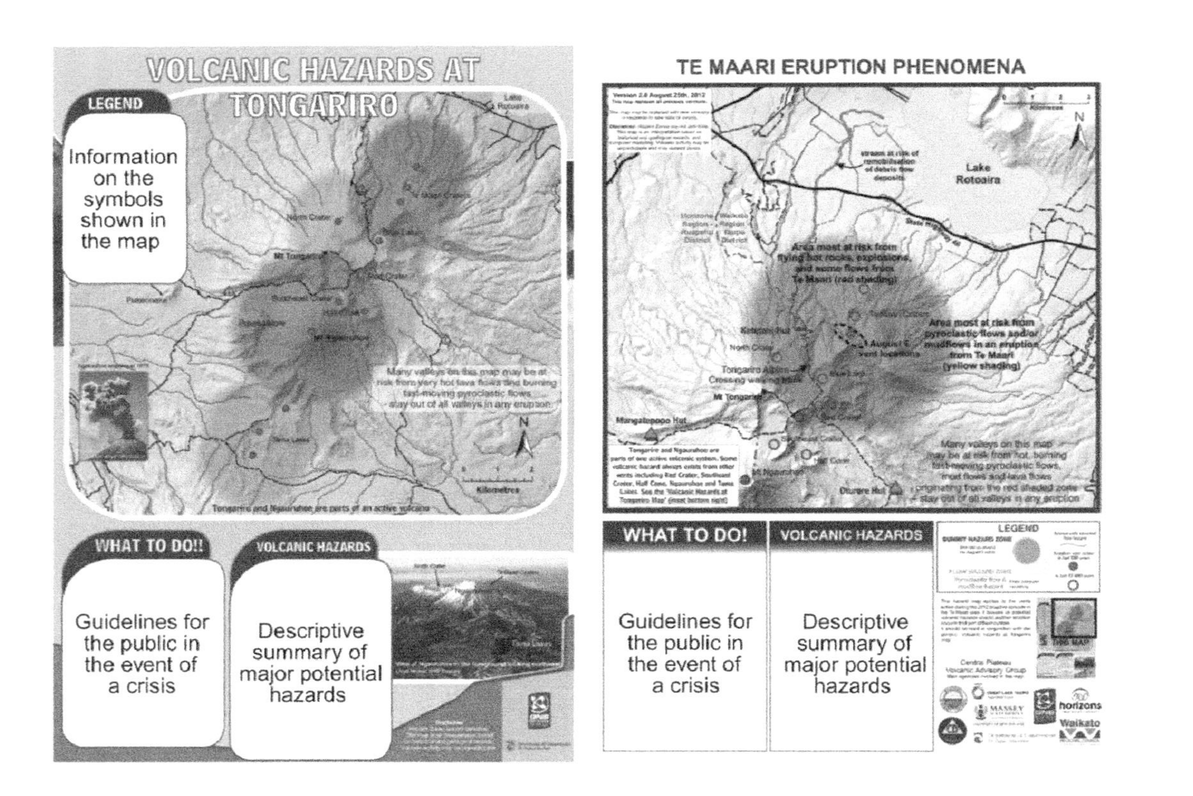

Figure 3.3. *Hazard-zone maps of the Tongariro volcanic center in the central North Island of New Zealand (GNS Science 2005, 2012; Leonard et al. 2014). For a color version of this figure, see www.iste.co.uk/lenat/hazards.zip*

COMMENT ON FIGURE 3.3.– *a) Traditional, long-term hazard-zone map illustrating hazards related to different types of direct and indirect eruptive activity. b) Short-term hazard-zone map oriented toward prevention during the 2012 "Te Maari" eruptive crisis, made in response to impending hazards. It proposes pragmatic measures for residents and tourists. We modified the maps as shown: small text areas replaced by general text.*

3.5.4. *Pioneering tests: Nevado del Ruiz (1985) and Mount Pelée (1985–1995)*

Classic hazard-zone maps are illustrated by the examples of Nevado del Ruiz (Thouret et al. 1987; Thouret, 1994) and Mount Pelée (Westercamp and Rançon 1983; Westercamp 1985; Rançon et al. 1995).

Published in a national newspaper a month before the tragic eruption of 1985, the map of potential hazards of Nevado del Ruiz had been drawn up based on deposits from the past eruptive activity of the volcano (see Figure 3.4). To this field evidence were added the debris avalanche and blast hypothesis (a directed lateral explosion producing a large-volume, particularly fast-moving type of surge) based on the events observed at Mount St. Helens in 1980 and on similar deposits found inside the avalanche scar on the northeast flank of Ruiz. The deposit map of the 1985 eruption (see Figure 3.5) shows that the paths taken by the lahars correspond to the predicted paths of the initial hazard map. However, the elliptical area of ash fallout, which was small and limited on November 13, 1985, oriented to the east-northeast (ENE) along the prevailing wind direction during the wet season and originating from a low-rise column, did not correspond to the concentric area predicted in October. In 1985, tephra dispersion models were poorly developed, and meteorological and statistical data were rarely incorporated into the models, except in the notable case of Mount St. Helens prior to its eruption. The third version of the Ruiz hazard-zone map, published by the Colombian Geological Survey in 2015, considered a fallout dispersion model (see Figure 3.6).

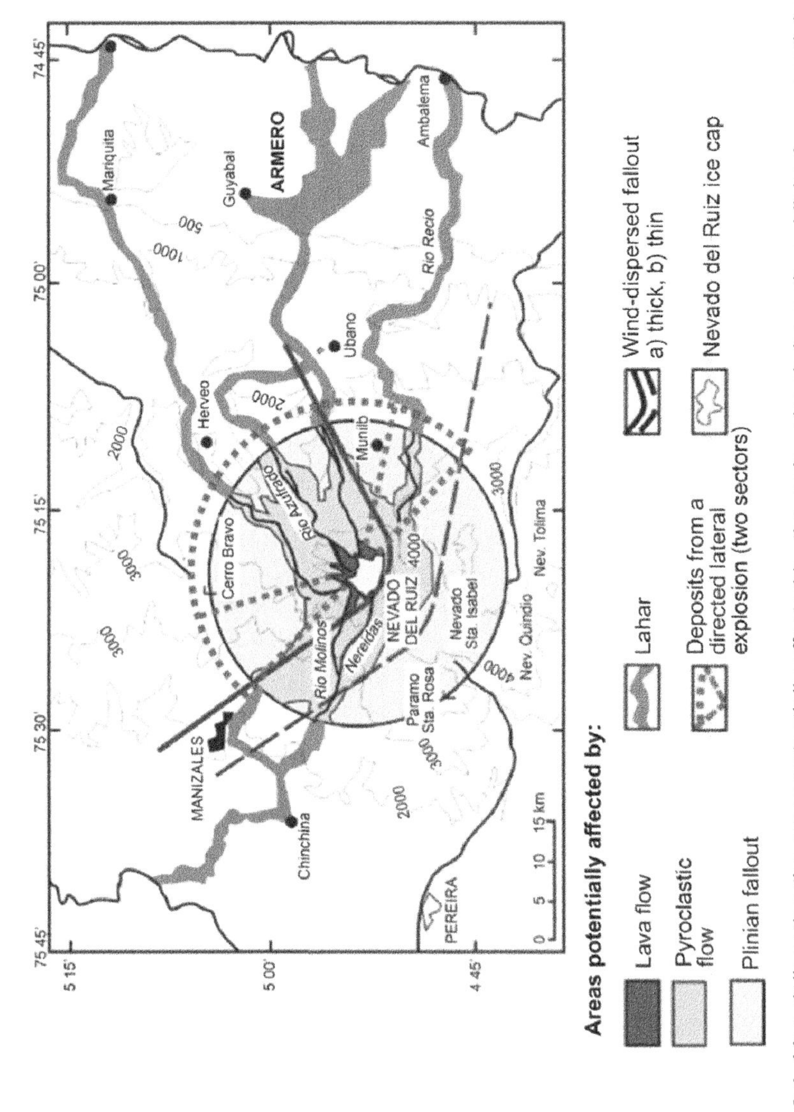

Figure 3.4. *Map delineating the areas potentially affected by the main volcanic hazards, published one month before the eruption in the national newspaper El Espectador. For a color version of this figure, see www.iste.co.uk/lenat/hazards.zip*

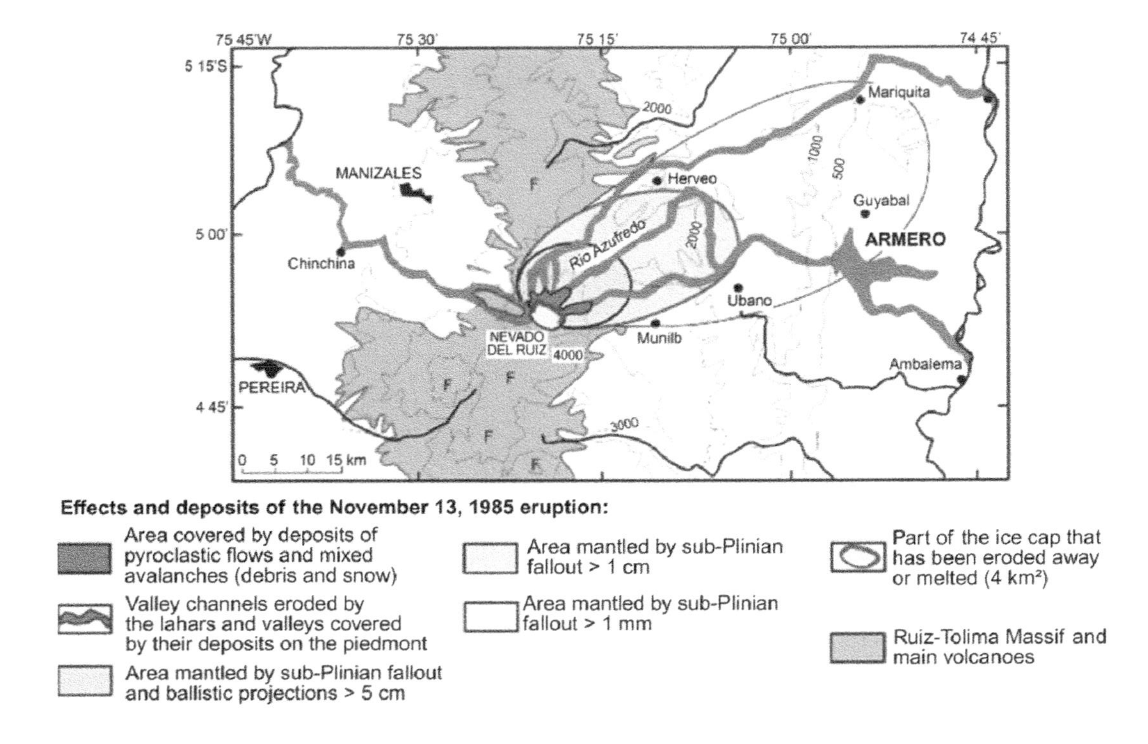

Figure 3.5. *Map of the deposits emplaced by the eruption of November 13, 1985. For a color version of this figure, see www.iste.co.uk/lenat/hazards.zip*

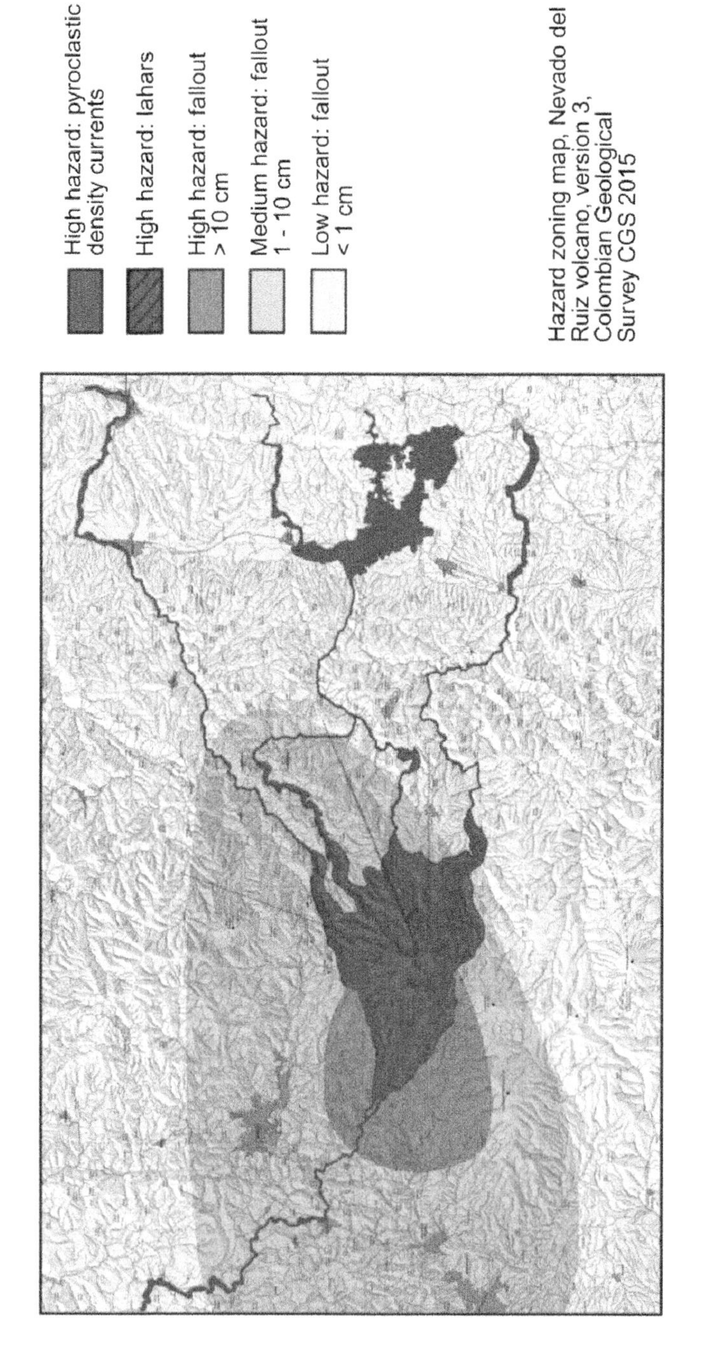

Figure 3.6. *Current geologic and deterministic hazard-zone map with a dispersion model for tephra fallout, revised by the Columbia Geological Survey (CGS) in 2015 (https://www2.sgc.gov.co/sgc/volcanes/VolcanNevadoRuiz/Paginas/Mapa-amenaza.aspx). For a color version of this figure, see www.iste.co.uk/lenat/hazards.zip*

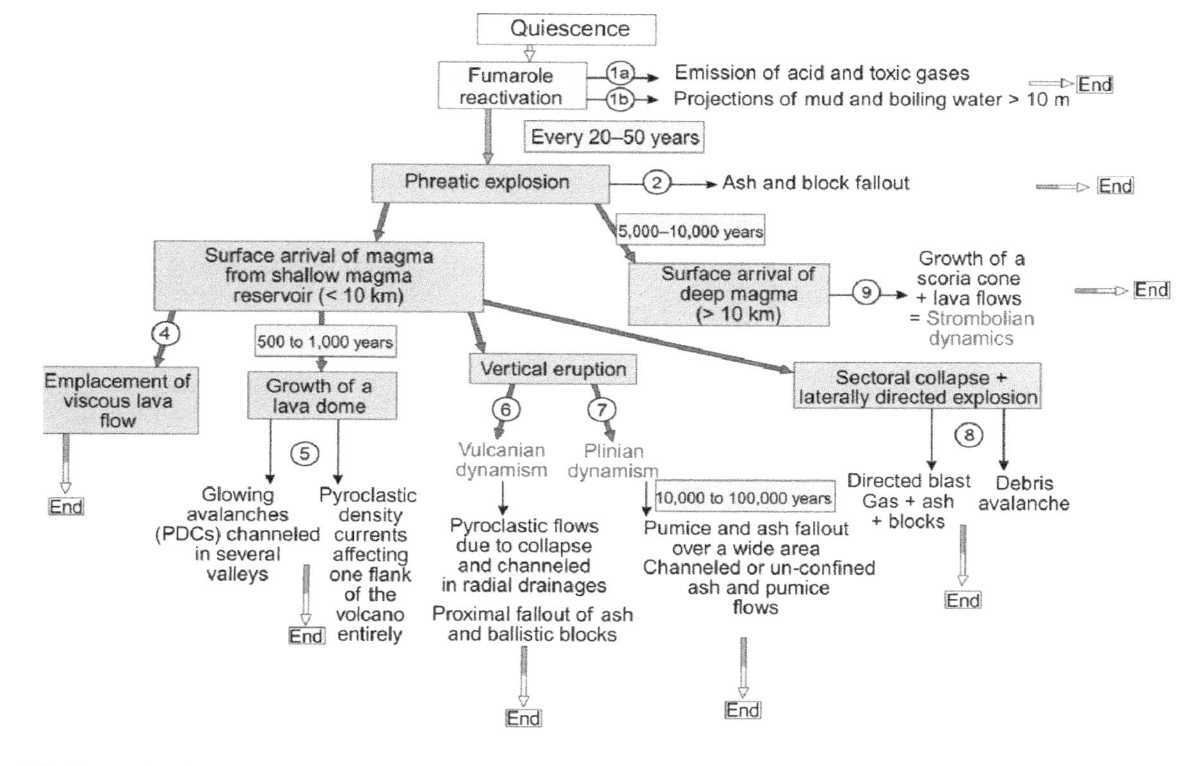

Figure 3.7. *Example of an eruptive scenario applied to the Soufrière of Guadeloupe (Komorowski et al. (2008); site of the Observatoire sismologique et volcanologique de la Guadeloupe, OSVG, IPGP (http://www.ipgp.fr/fr/ovsg/scenarios-eruptifs-possibles) (modified after Komorowski et al. (2008))*

The 1985 hazard-zone map of Mount Pelée by Westercamp and the eruptive scenarios by Rançon et al. (1995) highlight two issues during the 1980–1995 period.

– The precautionary principle was adopted for the delineation of certain hazards that are difficult to measure or reproduce experimentally, such as debris avalanches and PF. This principle led the authors to increase the probable runout distance of PF beyond the break-in-slope of the cone toward the volcano foothills.

– The hazard-zone map considers qualitatively the eruptive scenarios based on the occurrence and recurrence of eruptive phenomena observed on a volcano over a few hundred years. The method leads to a matrix of frequency and recurrence of all eruptive and associated phenomena, which the authors correlate in turn with the potential inundation area of the deposits. The concept of scenario gave rise to numerical and statistical developments in the early 2000s (Newhall and Hoblitt 2002). Since then, the Institute of Geophysics of Paris, in charge of the French Observatories, revised the scenarios of Mount Pelée eruptions, as was the case for the Soufrière of Guadeloupe in 2008 (see Figure 3.7).

A volcanic crisis and eruption scenario is a model usually depicted in a tree-like flow chart (see Figure 3.8) that considers the likely sequences of eruptive events and hazards over time and their probability of occurrence.

Each branch of the scenario leads from a conditional precursor event to an eventual outcome or to the emplaced deposits: for example, from an awakening to an explosive eruption to the Plinian eruptive style and then to an extensive pumice fallout. The delineation of the probable surface of this deposit stems from modeling taking into account the available meteorological data, the parameters of dispersion of the particles in the atmosphere and the physics of the eruptive column and plume.

An eruption scenario may lead to maps forming the layers of a GIS. It is based on knowledge of the volcano, its eruptive history, the frequency and magnitude of eruptions, the experience of volcanologists, monitoring data and elicitation or "expert judgment" methods conducted by experts who develop ranges of probabilities at each node in the event chain (see Figure 3.11).

The scenario allows, during a crisis, to guide the actions to be operated and the decisions to be made in terms of mitigating the effects of hazards.

Box 3.1. *Eruptive scenarios*

3.5.5. *Development of mapping techniques in the 1990s to 2000*

The example of the hazard maps of Merapi volcano (Thouret et al. 2020) illustrates the development of hazard-zone mapping techniques during the 1990s to 2000. At least 450,000 people live in areas classified as "hazard zones I and II", that is, exposed to a "very high" and "fairly high" risk of fallout, PFs and lahars. The authors delineated the potentially affected areas according to four scenarios ranked according to their magnitude and frequency. The first scenario, considered typical of Merapi's "Pelean" activity, includes frequent (2–8 years) small eruptions (volume $<4.10^6$ m^3) with block and ash flows and fallout, and less frequent (20–50 years) moderate eruptions (volume $4–10.10^6$ m^3) with larger PFs and lahars. The second scenario considers larger eruptions (VEI 4, volume $>10^7$ m^3), such as the effusive and pyroclastic episode that occurred in 1930–1931. The other scenarios derive from the large Plinian eruption of 1872, whose recurrence is secular, and from catastrophic eruptions that could trigger the collapse of the summit dome, or even of the southern flank, inducing a blast, and then voluminous lahars, whose examples are probably prehistoric and whose recurrence is low.

Following a different approach, the Merapi Volcanological Observatory (MVO or BPPTKG) published a revised map in 2011 that delineates three "danger zones" related to future eruptions at Merapi. This delineation, based on geological data, extent of the 2010 deposits, and simulations using LAHARZ (a semi-empirical and statistical code developed by Schilling (1998, 2014)), remains deterministic.

The integration of numerical simulations, using models such as VolcFlow and Titan2D or FLO-2D, may provide more relevant information (Charbonnier et al. 2013; Kelfoun et al. 2017). Indeed, analysis of the 2010 Merapi eruption (detailed field study of deposits) (Cronin et al. 2013; Komorowski et al. 2013) and simulation of the surface area, thickness, volume and trajectories of dilute, unconfined surges compared to those of dense, channeled flows (Kelfoun et al. 2017) demonstrated the need to develop more sophisticated scenarios. These methods were complemented by the habitat damage inventory that led to the mapping of dynamic pressure impacts from PFs and lahars (Jenkins et al. 2013, 2015). In addition, morphometric analysis of drainage channels aimed at reconstructing and then predicting the overbank and avulsion or re-channelization of PFs (Lube et al. 2011; Charbonnier et al. 2013) and lahars (De Bélizal et al. 2013; Solikhin et al. 2015).

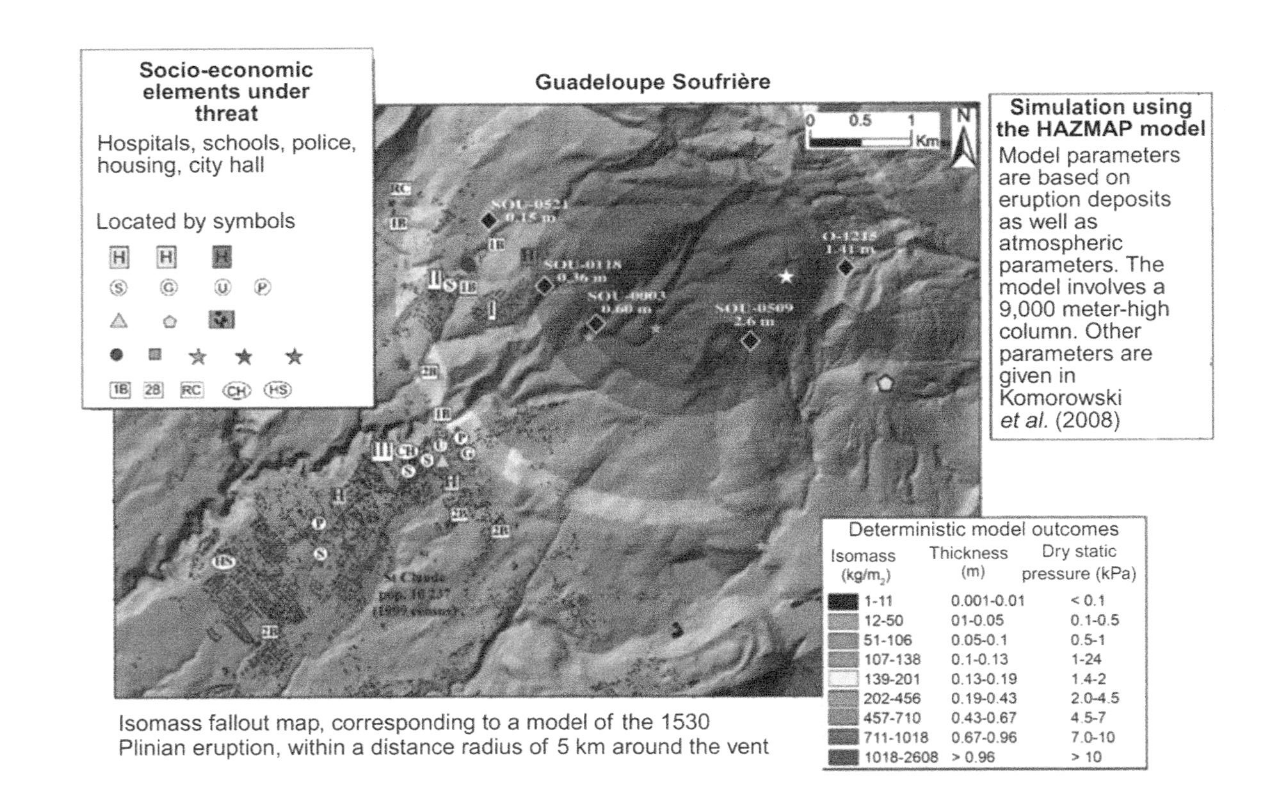

Figure 3.8. *Deterministic hazard modeling, using the HAZMAP model, based on the case of the 1530 ce Plinian eruption at La Soufrière in Guadeloupe (Komorowski et al. 2008). For a color version of this figure, see www.iste.co.uk/lenat/hazards.zip*

3.6. New approaches to modeling and quantitative analysis

3.6.1. *Evolution of delineation methods: DTM, GIS and digital codes*

A new approach to hazard-zone delineation has emerged since the 2000s, following eruptions with disastrous consequences such as Pinatubo in 1991 (catastrophic lahars until 2004), Soufrière Hills (1995–2010), Sinabung since 2010, Calbuco in 2015, Agung in 2017 and Fuego in 2018. It consists of a post-eruption damage inventory, similar to the method developed for post-earthquake damage analysis (Jenkins et al. 2013, 2015). It leads to the probabilistic analysis of potential damage. It requires the identification of deposits, inventory and statistical analysis of damage, and then simulation of potential footprints and areas covered, for example, by fallout, lahars and PFs. The resulting hazard-zone maps present at least a distribution of the exposed elements (buildings, bridges and roads, water, electricity and transport networks, and health system) according to a chosen or probable impact scenario. This is the case in Figure 3.8, depicting a deterministic scenario based on the effects of a given historical eruption (a Plinian eruption such as the 1530 eruption) that includes the elements exposed to the potential impacts of the eruption around the Guadeloupe Soufrière (Komorowski et al. 2008).

Delineation methods now exploit an array of tools and complementary information: DTM, GIS, ground and space-based remote sensing, numerical codes and inventory of assets (habitat, infrastructure, networks, heritage, governance, including the prevention and emergency system), and even modeling of losses in the event of a disaster. Among the main characteristics of these modern methods, we note the following points:

– Flow simulation models are very sensitive to the resolution and accuracy of DTMs. DTMs with accuracy suitable for small-scale modeling can be computed using a variety of methods: stereophotogrammetry using sub-meter resolution imagery acquired by satellites, aircraft or drones, and high-spatial resolution LIDAR surveys. A challenge may be the frequent updating of a DTM when the topography has been modified by previous phases of an eruption.

– The various information can be managed digitally (GIS or a geomatic platform). In order to exploit the hazard-zone maps in terms of risk, structural (buildings, roads and networks) and functional (exposure indices and socio-economic vulnerability indicators) spatial data will be used. The evaluation of all exposed elements can then be integrated into a probabilistic delineation of the impacts of hazards. A geomatic platform allows the integration of all the territorial and socio-economic databases necessary for the delineation of hazard zones and, beyond, for emergency measures and land use planning in the longer term. Examples have been developed for Popocatépetl volcano at SUNY, Buffalo, USA (Sheridan et al. 2001), as well as for the island of Montserrat (Calder et al. 2015) and Naples (Lirer et al. 2010).

– Researchers often use space- and ground-based remote sensing for volcano monitoring and hazard assessment. In the case of cloud cover, radar remote sensing can supplement visible and infrared imagery, as illustrated in October–November 2010 during the Merapi volcano eruption in Indonesia (Pallister et al. 2013).

– Numerical models aim at reproducing or forecasting the area likely to be affected, the paths and the velocity of eruptive phenomena, as well as the extent and thickness of their deposits. After the introduction of the concept of energy cone (theoretical cone covering the volcano up to the slope break between the cone and its foothills) and the ratio of edifice height/distance reached by flows by Malin and Sheridan (1982), one of the first semi-empirical codes was LAHARZ, created by Schilling (1998, 2014) and implemented by Iverson et al. (1998) to simulate lahar flood-prone zones in the radial valleys of Mount Rainier. The numerical codes now most commonly used for hazard assessment are Tephra2, Ash3D (Mastin et al. 2013) for tephra fallout; Eject for ballistics; Titan2D, TITAN2F, FLO-2D and VolcFlow for PFs and lahars; DOWNFLOW, FLOWGO, MAGFLOW, LavaSIM and GPSUPH for lava flows (see Figure 3.9) (Harris and Rowland 2015); and VolcFlow for debris avalanches. Hazard zone delineation obtained from numerical simulations has been validated at Soufriere Hills (Wadge et al. 1998) and Vesuvius (Zuccaro et al. 2008), while DOWNFLOW helped Favalli et al. (2009) define the spatiotemporal probability of building damage at Etna. However, none of the models can take into account all the static and dynamic parameters of the simulated phenomena. Each has its own advantages and limitations. For this reason, researchers conduct comparative studies to illustrate their strengths and weaknesses. Figures 3.9 and 3.10[2] provide examples of comparison between simulation model results.

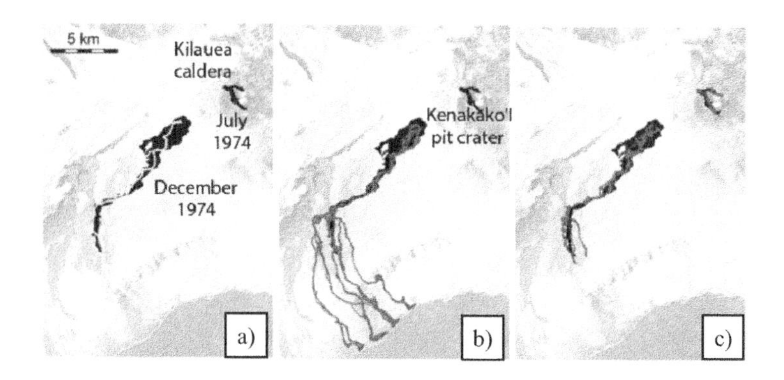

Figure 3.9. *Comparison between lava flow models simulating the July and December 1974 lava flows at Kilauea (Harris and Rowland 2015). For a color version of this figure, see www.iste.co.uk/lenat/hazards.zip*

COMMENT ON FIGURE 3.9.– *a) Flow path of the December 1974 lava flow along the steepest slope. b) Multiple potential flow paths (in red) from the stochastic DOWNFLOW model. c) Simulation of potential runouts and flow paths identified by the model in figure (b) with the thermo-rheological FLOWGO model (Harris and Rowland 2015).*

Figure 3.10 shows that by imposing validation metrics based on a Bayesian approach, one can quantitatively evaluate the performance of each numerical model to reproduce a past event. These results not only serve to improve the accuracy of the models but also to transform them into a robust tool for delineating hazard-prone areas.

2 The "Jaccard" index is the ratio of the area of intersection of observed and simulated PFs to the area of their union. The "sensitivity" index is the ratio of the intersection area of observed and simulated PFs to the area of simulated PFs only, while the "specificity" index is the ratio of the intersection area of observed and simulated PFs to the area of observed PFs only. The two values "Positive Predictive" and "Negative Predictive" are Bayesian indices that measure the confidence of the model to simulate (positive value) or not to simulate (negative value) the area of observed PFs, in relation to the potential PF hazard area (in pink on the map). The values of the validation indices in the table vary from 0 to 1, with 1 corresponding to a total reproducibility of the simulations of the area covered by the real PF. The quality of the validation appears in color in the table: from limited, in orange, to good, in yellow, to very good, in white (Charbonnier et al. 2018).

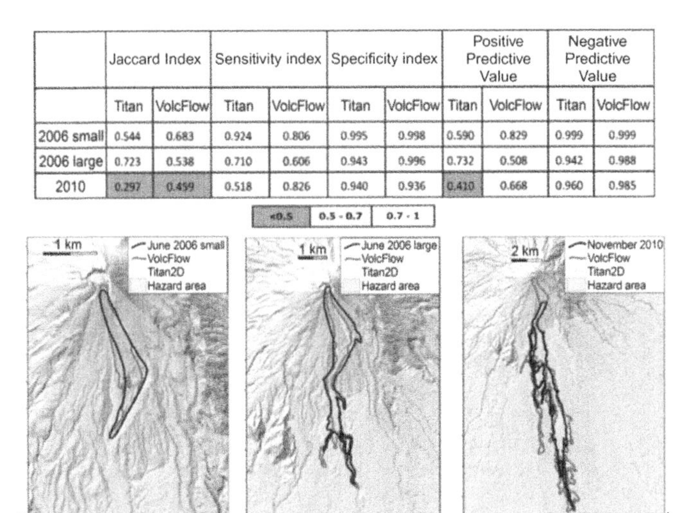

	Jaccard Index		Sensitivity index		Specificity index		Positive Predictive Value		Negative Predictive Value	
	Titan	VolcFlow	Titan	VolcFlow	Titan	VolcFlow	Titan	VolcFlow	Titan	VolcFlow
2006 small	0.544	0.683	0.924	0.806	0.995	0.998	0.590	0.829	0.999	0.999
2006 large	0.723	0.538	0.710	0.606	0.943	0.996	0.732	0.508	0.942	0.988
2010	0.297	0.459	0.518	0.826	0.940	0.936	0.410	0.668	0.960	0.985

Figure 3.10. *Three maps of dense channelized pyroclastic flow (PF, in black) deposits and associated unchanneled overflows at Merapi in 2006 ("2006 small" = PF of June 8, 2006, "2006 large" = PF of June 14, 2006) and 2010 (PF of November 4–5, 2010), from both VolcFlow (blue) and Titan2D (yellow) models, and table of validation metrics used to evaluate the model performance. For a color version of this figure, see www.iste.co.uk/lenat/hazards.zip*

3.6.2. *The statistical, probabilistic and evolutionary representation of delineation of hazard zones*

Since 2004, a Bayesian approach (statistics adapted to small sample sets such as the eruption history of a volcano) has prevailed in the evaluation of volcanic hazards, either for long-term delineation or within observatories that are confronted with the management of a prolonged or repeated crisis. This approach has found expression in the Bayesian Event Tree (BET) software (Marzocchi et al. 2008, 2010; Selva et al. 2010; Sandri et al. 2012), which is available in two modes, VH (Volcanic Hazard) and EF (Eruption Forecasting). The procedure calculates the probability of any long-term hazard by taking into account the intrinsic stochastic variability of volcanic eruptions and the lack of knowledge about eruptive processes. The code incorporates the results of numerical models for various hazards, as well as data from chrono-stratigraphic analysis of the history of the volcano under consideration, or from nearby and known neighboring or analogous volcanoes, or even the signals obtained from instrumental and phenomenological monitoring. The researchers of the Volcanological

Observatory and the University of Naples have acquired robust BET research outcomes on Vesuvius (Zuccaro et al. 2008).

The BET code provides a wide range of spatial and temporal probabilities of various hazards. It is able to handle all eruptive styles at the same time, ranking and weighting each one with its own probability of occurrence. For example, an event tree depicts four PF scenarios associated with the growth of the Soufriere Hills dome in Montserrat (see Figure 3.11) (Sparks et al. 2011, their Figure 11.9). Such an approach has allowed these authors to discuss the probabilities of occurrence of different eruptive styles and how phreatic, phreato-magmatic, or magmatic and seismo-volcanic events will occur in sequence after a volcano awakens (Selva et al. 2010). When researchers do not fully understand volcanic processes, probabilistic estimates may be entirely empirical, that is, volcanologists use observations of past and current activity and the assumption that future activity will mimic that of the past or the current trend. When experts understand eruptive processes, they can estimate probabilities from a theoretical model, using statistical models or numerical simulations of the various phenomena. Decision trees used during volcanic crises can help volcanologists revise their hazard analysis in real time and authorities and the public compare the volcanic risk with the one they experience in their daily life. These decision trees highlight the inherently probabilistic nature of volcanic forecasts.

3.6.3. *Large-scale delineation of hazard zones*

Sometimes, the assessment focuses on a hazard at a regional or continental scale. For example, Jenkins et al. (2012) have addressed the case of tephra fallout across Southeast Asia and Australasia. From the catalog of active volcanoes, the author compiled the potential hazard of a 1-cm-thick ash fallout in a region or city in a year. This allows an analysis of the main risks for cities exposed to tephra fallout hazards, as well as the distinction between the role of hazard, vulnerability and impacts associated with such phenomenon.

Another example is the hazards associated with the Phlegraean Fields and Vesuvius in the densely populated Campanian region. Lirer et al. (2010) based their assessment on decades of research and observations and multiple previous studies. The integrated assessment considers all volcanic and seismic hazards and risks in the region.

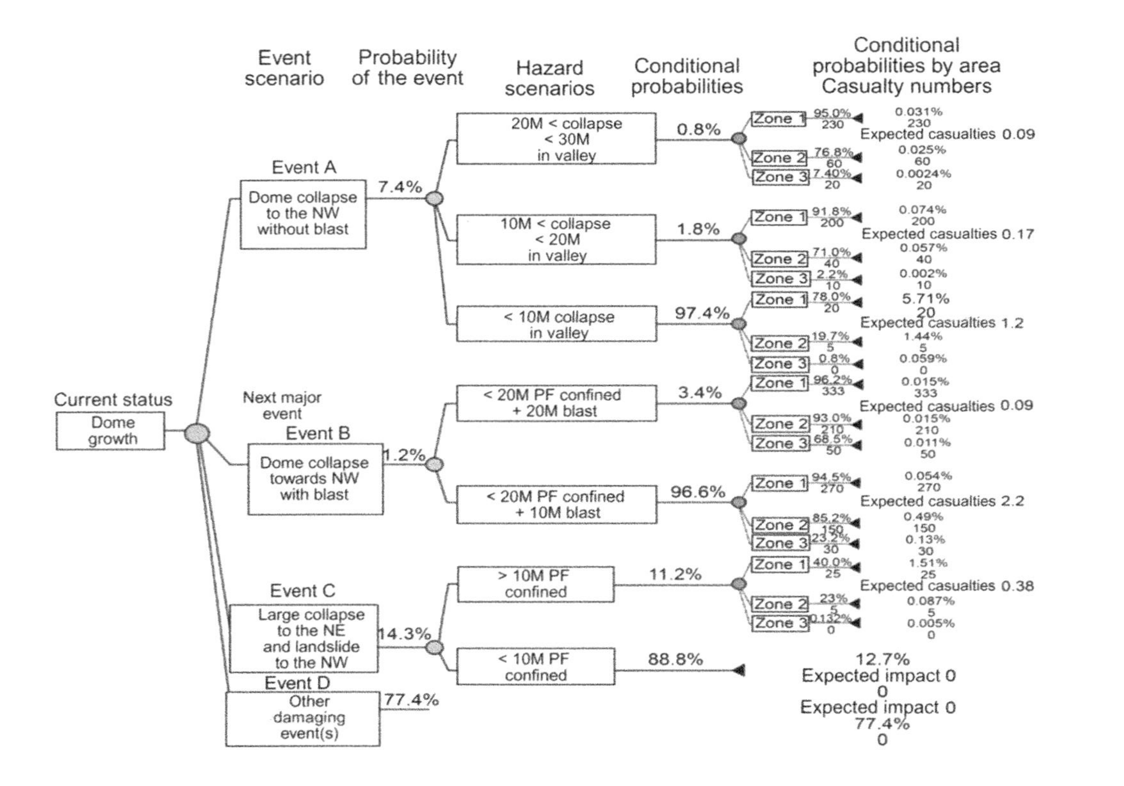

Figure 3.11. *Probabilistic event tree, applied to the cases of dome collapse and a blast (directed lateral explosion) of one of the domes of Soufrière Hills volcano, island of Montserrat (Sparks et al. (2011), Figure 11.9, p. 381)*

COMMENT ON FIGURE 3.11.– *This is a probabilistic event tree, applied to the case of dome collapse and a blast (directed lateral explosion) of one of the domes of Soufrière Hills volcano. The authors predicted three scenarios, A–C, for three populated areas and classified them according to volumes (M for million m^3). A fourth scenario, D, would not affect the valley under study. The probabilities assigned to each branch result from an elicitation procedure and expert judgment (Aspinall and Blong 2015). The number of potential victims, based on experience, depends on the number of residents in each area and the type of hazard. The values (in %) show a range of error that represents the uncertainty of the method. Each expected casualty number indicates a statistical value for a particular scenario selected. The set of expected values shows that the most likely and least intense eruptive events are the ones that present more risk, while the most intense events may cause more casualties but are proportionally much less frequent.*

3.7. Conclusion

The delineation of volcanic hazards remains complex and fraught with uncertainty but quickly evolves thanks to major advances in volcanology, new tools available and recent advances in eruption forecasting. This is important because the worldwide value of the stakes increases along with economic and demographic development.

A crucial point is the communication of risk via hazard-zone maps. We can express the argument of Thompson et al. (2018): "Maps communicate more than meets the eye"; the map is a communication tool that goes beyond what the eye is capable of perceiving. Reading a map actually depends on the knowledge and experience of the reader interpreting the visual data. However, this complex exercise can produce effects contrary to those intended. Hazard maps are accessible to different audiences during crises. However, they may have difficulties in understanding the input parameters of the delineation and may not be familiar with the results of the models used. It is therefore important to account for these handicaps and to accompany the communication of the maps by specific exchanges and debates according to the public (inhabitants, decision-makers, etc.).

3.8. References

Aspinall, W. and Blong, R. (2015). Volcanic risk assessment. In *The Encyclopedia of Volcanoes*, Sigurdsson, H., Houghton, B.F., McNutt, S.R., Rymer, H., Stix, J. (eds). Academic Press, London.

Auker, M.R., Sparks, R.S.J., Siebert, L., Crosweller, H.S., Ewert, J. (2013). A statistical analysis of the global historical volcanic fatalities record. *Journal of Applied Volcanology*, 2(1), 2.

Blong, R. (2000). Volcanic hazards and risk management. In *The Encyclopedia of Volcanoes*, Sigurdsson, H., Houghton, B.F., McNutt, S.R., Rymer, H., Stix, J. (eds). Academic Press, San Diego.

Brown, S.K., Jenkins, S.F., Sparks, R.S.J., Odbert, H., Auker, M.R. (2017). Volcanic fatalities database: Analysis of volcanic threat with distance and victim classification. *Journal of Applied Volcanology*, 6(1), 1–20.

Calder, E.S., Wagner, K., Ogburn, S.E. (2015). Volcanic hazard maps. In *Global Volcanic Hazards and Risk*, Loughlin, S.C., Sparks, R.S.J., Brown, S.K., Jenkins, S.F., Vye-Brown, C. (eds). Cambridge University Press, Cambridge.

Charbonnier, S.J., Germa, A., Connor, C.B., Gertisser, R., Preece, K., Komorowski, J.C. et al. (2013). Evaluation of the impact of the 2010 pyroclastic density currents at Merapi volcano from high-resolution satellite imagery, field investigations and numerical simulations. *Journal of Volcanology and Geothermal Research*, 261, 295–315.

Charbonnier, S.J., Macorps, E., Connor, C.B., Connor, L.J., Richardson, J.A. (2018). How to correctly evaluate the performance of volcanic mass flow models used for hazard assessment? In *Hazard and Risk Mapping. The Arequipa El Misti Case Study and Other Threatened Cities*, Thouret, J.C. (ed.). Presses Universitaires Blaise Pascal, Clermont-Ferrand.

Chester, D.K., Degg, M., Duncan, A.M., Guest, J.E. (2000). The increasing exposure of cities to the effects of volcanic eruptions: A global survey. *Environmental Hazards*, 2(3), 89–103.

Cronin, S.J., Lube, G., Dayudi, D.S., Sumarti, S., Subrandiyo, S. (2013). Insights into the October-November 2010 Gunung Merapi eruption (Central Java, Indonesia) from the stratigraphy, volume and characteristics of its pyroclastic deposits. *Journal of Volcanology and Geothermal Research*, 261, 244–259.

D'Ercole, R. and Metzger, P. (2004). La vulnerabilidad del distrito metropolitano de Quito [Online]. Available at: https://biblio.flacsoandes.edu.ec/shared/biblio_view.php?bibid=15995&tab=opac.

De Bélizal, E., Lavigne, F., Hadmoko, D.S., Degeai, J.P., Dipayana, G.A., Mutaqin, B.W. et al. (2013). Rain-triggered lahars following the 2010 eruption of Merapi volcano, Indonesia: A major risk. *Journal of Volcanology and Geothermal Research*, 261, 330–347.

Favalli, M., Tarquini, S., Fomaciai, A., Boschi, E. (2009). A new approach to risk assessment of lava flow at Mount Etna. *Geology*, 37(12), 1111–1114.

GNS Science (2005). Volcanic hazards at Tongariro [Online]. Available at: https://www.doc.govt.nz/globalassets/documents/parks-and-recreation/tracks-and-walks/tongariro-taupo/tongariro-poster-a4.pdf [Accessed December 1, 2021].

GNS Science (2012). Te Maari eruption phenomena map [Online]. Available at: https://www.gns.cri.nz/Home/Learning/Science-Topics/Volcanoes/New-Zealand-Volcanoes/Tongariro/Te-Maari-eruption-phenomena-map [Accessed December 2, 2021].

Harris, A.J.L. and Rowland, S.K. (2015). Flowgo 2012: An updated framework for thermorheological simulations of channel-contained lava. In *Geophysical Monograph Series*, Volume 208, Carey, R., Cayol, V., Poland, M., Weis, D. (eds). doi.org/10.1002/9781118872079.ch21.

Iverson, R.M., Schilling, S.P., Vallance, J.W. (1998). Objective delineation of lahar-inundation hazard zones. *GSA Bulletin*, 110(8), 972–984.

Jenkins, S., Magill, C., McAneney, J., Blong, R. (2012). Regional ash fall hazard I: A probabilistic assessment methodology. *Bulletin of Volcanology*, 74(7), 1699–1712.

Jenkins, S., Komorowski, J.C., Baxter, P.J., Spence, R., Picquout, A., Lavigne, F., Surono, S. (2013). The Merapi 2010 eruption: An interdisciplinary impact assessment methodology for studying pyroclastic density current dynamics. *Journal of Volcanology and Geothermal Research*, 261, 316–329.

Jenkins, S.F., Phillips, J.C., Price, R., Feloy, K., Baxter, P.J., Hadmoko, D.S., de Bélizal, E. (2015). Developing building-damage scales for lahars: Application to Merapi volcano, Indonesia. *Bulletin of Volcanology*, 77(9), 1–17.

Kelfoun, K., Gueugneau, V., Komorowski, J.C., Aisyah, N., Cholik, N., Merciecca, C. (2017). Simulation of block-and-ash flows and ash-cloud surges of the 2010 eruption of Merapi volcano with a two-layer model. *Journal of Geophysical Research: Solid Earth*, 122(6), 4277–4292.

Komorowski, J.-C., Legendre, Y., Caron, B., Boudon, G. (2008). Reconstruction and analysis of sub-plinian tephra dispersal during the 1530 A.D. Soufrière (Guadeloupe) eruption: Implications for scenario definition and hazards assessment. *Journal of Volcanology and Geothermal Research*, 178(3), 491–515.

Komorowski, J.-C., Jenkins, S., Baxter, P.J., Picquout, A., Lavigne, F., Charbonnier, S. et al. (2013). Paroxysmal dome explosion during the Merapi 2010 eruption: Processes and facies relationships of associated high-energy pyroclastic density currents. *Journal of Volcanology and Geothermal Research*, 261, 260–294.

Leonard, G.S., Stewart, C., Wilson, T.M., Procter, J.N., Scott, B.J., Keys, H.J. et al. (2014). Integrating multidisciplinary science, modelling and impact data into evolving, syn-event volcanic hazard mapping and communication: A case study from the 2012 Tongariro eruption crisis, New Zealand. *Journal of Volcanology and Geothermal Research*, 286, 208–232.

Lirer, L., Petrosino, P., Alberico, I. (2010). Hazard and risk assessment in a complex multi-source volcanic area: The example of the Campania Region, Italy. *Bulletin of Volcanology*, 72(4), 411–429.

Lube, G., Cronin, S.J., Thouret, J.C., Surono, S. (2011). Kinematic characteristics of pyroclastic density currents at Merapi and controls on their avulsion from natural and engineered channels. *Bulletin of the Geological Society of America*, 123(5), 1127–1140.

Major, J.J. and Lara, L.E. (2013). Overview of Chaitén Volcano, Chile, and its 2008–2009 eruption. *Andean. Geol.*, 40, 2, 196–215.

Malin, M.C. and Sheridan, M.F. (1982). Computer-assisted mapping of pyroclastic surges. *Science*, 217(4560), 637–640.

Marzocchi, W., Sandri, L., Selva, J. (2008). BET_EF: A probabilistic tool for long-and short-term eruption forecasting. *Bulletin of Volcanology*, 70(5), 623–632.

Marzocchi, W., Sandri, L., Selva, J. (2010). BET_VH: A probabilistic tool for long-term volcanic hazard assessment. *Bulletin of Volcanology*, 72(6), 705–716.

Mastin, B.L.G., Randall, M.J., Schwaiger, H.F., Denlinger, R.P. (2013). User's guide and reference to Ash3D: A 3-D Eulerian atmospheric tephra transport and deposition model. *U.S. Geol. Survey Open-File Report*, 48. doi.org/10.3133/ofr20131122.

Newhall, C. and Hoblitt, R. (2002). Constructing event trees for volcanic crises. *Bulletin of Volcanology*, 64(1), 3–20.

Newhall, C.G. and Punongbayan, R. (1996). *Fire and Mud. Eruptions and Lahars of Mount Pinatubo, The Philippines*. University of Washington, Seattle.

Pallister, J.S., Schneider, D.J., Griswold, J.P., Keeler, R.H., Burton, W.C., Noyles, C. et al. (2013). Merapi 2010 eruption – Chronology and extrusion rates monitored with satellite radar and used in eruption forecasting. *Journal of Volcanology and Geothermal Research*, 261, 144–152.

Pierson, T.C. and Major, J.J. (2014). Hydrogeomorphic effects of explosive volcanic eruptions on drainage basins. *Annual Review of Earth and Planetary Sciences*, 42(1), 469–507.

Pierson, T.C., Janda, R.J., Thouret, J.C., Borrero, C.A. (1990). Perturbation and melting of snow and ice by the 13 November 1985 eruption of Nevado del Ruiz, Colombia, and consequent mobilization, flow and deposition of lahars. *Journal of Volcanology and Geothermal Research*, 41(1/4), 17–66.

Rançon, J.P., Repusseau, P., Sedan, O. (1995). Base de données sur la phénoménologie des éruptions de la Montagne Pelée. Deuxième phase : développement de la base et animation multimedia de la crise de 1889-1905. DIPCN/Antilles-Caraïbe. Rapport du BRGM R, 38, 243, référence P04232054. DPPR, Ministère de l'Environnement, Paris.

Sandri, L., Jolly, G., Lindsay, J., Howe, T., Marzocchi, W. (2012). Combining long- and short-term probabilistic volcanic hazard assessment with cost-benefit analysis to support decision making in a volcanic crisis from the Auckland Volcanic Field, New Zealand. *Bulletin of Volcanology*, 74(3), 705–723.

Schilling, S.P. (1998). Laharz: GIS programs for automated mapping of lahar-inundation hazard zones. Open-File Report, U.S. Geological Survey, Reston.

Schilling, S.P. (2014). *Laharz_py: GIS Tools for Automated Mapping of Lahar Inundation Hazard Zones*. U.S. Geological Survey, Reston.

Scott, K.M., Vallance, J.W., Kerle, N., Macías, J.L., Strauch, W., Devoli, G. (2005). Catastrophic precipitation-triggered lahar at Casita volcano, Nicaragua: Occurrence, bulking and transformation. *Earth Surface Processes and Landforms*, 30(1), 59–79.

Self, S. (2006). The effects and consequences of very large explosive volcanic eruptions. *Philosophical Transactions of the Royal Society A: Mathematical, Physical and Engineering Sciences*, 364(1845), 2073–2097.

Selva, J., Costa, A., Marzocchi, W., Sandri, L. (2010). BET_VH: Exploring the influence of natural uncertainties on long-term hazard from tephra fallout at Campi Flegrei (Italy). *Bulletin of Volcanology*, 72(6), 717–733.

Sheridan, M.F., Hubbard, B., Bursik, M.I., Abrams, M., Siebe, C., Macías, J.L., Delgado, H. (2001). Gauging short-term volcanic hazards at Popocatepetl. *Eos*, 82(16), 185–189.

Solikhin, A., Thouret, J.-C., Liew, S.C., Gupta, A., Sayudi, D.S., Oehler, J.-F., Kassouk, Z. (2015). High-spatial-resolution imagery helps map deposits of the large (VEI 4) 2010 Merapi Volcano eruption and their impact. *Bulletin of Volcanology*, 77(3), 20.

Sparks, R.S., Aspinall, W.P., Crosweller, H.S., Hincks, T.K. (2011). Risk and uncertainty assessment of volcanic hazards. In *Risk and Uncertainty Assessment for Natural Hazards*, Rougier, J., Sparks, R.J.S., Hill, L. (eds). Cambridge University Press, Cambridge.

Stieltjes, L. (2004). Analyse du risque volcanique : état de l'art sur l'aléa volcanique [Online]. Available at: http://infoterre.brgm.fr/rapports/RP-53006-FR.pdf.

Thompson, M.A., Lindsay, J.M., Leonard, G.S. (2018). More than meets the eye: Volcanic hazard map design and visual communication. In *Observing the Volcano World. Volcano Crisis Communication*, Fearnley, C.J., Bird, D.K., Haynes, K., McGuire, W.J., Jolly, G. (eds). Springer, Berlin.

Thouret, J.-C. (1994). Méthodes de zonage des menaces et des risques volcaniques. In *Le Volcanisme*, Bourdier, J.L., Boivin, P., Gourgaud, A., Camus, G., Vincent, P.M., Lénat, J.F. (eds). BRGM, Orléans.

Thouret, J.-C. (2004). Geomorphic processes and hazards on volcanic mountains. In *Mountain Geomorphology*, 1st edition, Owens, P.N., Slaymaker, O. (eds). Arnold, London.

Thouret, J.-C., Janda, R., Pierson, T., Calvache, M., Cendrero, A. (1987). L'éruption du 13 novembre 1985 au Nevado El Ruiz (Cordillère Centrale, Colombie) : interactions entre le dynamisme éruptif, la fusion glaciaire et la genèse d'écoulements volcano-glaciaires. *Comptes Rendus de l'Académie Des Sciences. Série 2, Mécanique, Physique, Chimie, Sciences de l'univers, Sciences de La Terre*, 305(6), 505–509.

Thouret, J.-C., Antoine, S., Magill, C., Ollier, C. (2020). Lahars and debris flows: Characteristics and impacts. *Earth-Science Reviews*, 201, 479–502.

Tilling, R.I. (2005). Volcanic hazards. In *Volcanoes and the Environment*, Marti, J., Ernst, G.J. (eds). Cambridge University Press, Cambridge.

Wadge, G., Jackson, P., Bower, S.M., Woods, A.W., Calder, E. (1998). Computer simulations of pyroclastic flows from dome collapse. *Geophysical Research Letters*, 25(19), 3677–3680.

Westercamp, D. (1985). La prévision générale des risques volcaniques. Méthodologie. Application à la Montagne Pelée, Martinique. Rapport BRGM no. 85 SGN 421 IRG, Orléans.

Westercamp, D. and Rançon, P. (1983). *Carte géologique de la montagne Pelée, Martinique (1/20 000)*. BRGM, Orléans.

Zuccaro, G., Cacace, F., Spence, R.J.S., Baxter, P.J. (2008). Impact of explosive eruption scenarios at Vesuvius. *Journal of Volcanology and Geothermal Research*, 178(3), 416–453.

4

History of Volcanic Monitoring and Development of Methods

Jean-François LÉNAT
Laboratoire Magmas et Volcans, CNRS, IRD, OPGC, Université Clermont Auvergne, Clermont-Ferrand, France

4.1. Qualitative observation

Volcanoes and their activity have always captivated human interest. Before arousing scientific curiosity, they inspired fear and were associated with supernatural powers. Cults dedicated to these powers can be found in almost every volcanic region of the world. There is a large literature on the interactions between volcanoes and societies (Fisher et al. 1997; de Boer and Sanders 2002; Sheets 2015).

Beyond the supernatural interpretations, some ancient narratives contain descriptions and observations of definite scientific interest. Moreover, before the development of instrumental monitoring (late 19th–early 20th centuries), these qualitative data represented the only source of information on eruptive activity. It should be noted that, even in our era of sophisticated instrumental monitoring, visual observations remain indispensable for the interpretation of many signals.

The oldest useful description of eruptive activity is that of Pliny the Younger in his letter to Tacitus where he reports the eruption of Vesuvius in 79 CE and the death of his uncle Pliny the Elder. He describes the eruptive plume of Vesuvius as follows:

Hazards and Monitoring of Volcanic Activity 1,
coordinated by Jean-François LÉNAT. © ISTE Ltd 2022.

> A cloud was formed (one could not see from afar which mountain it came from, it was then known that it was Vesuvius), having the appearance and shape of a tree and making one think especially of a pine tree. For after having risen in the manner of a very elongated trunk, it unfolded like branches, having been at first, I suppose, carried upwards by the column of air at the moment when it had originated, then this column having fallen back, abandoned to itself or yielding to its own weight, it vanished while widening; in places it was of a brilliant white, elsewhere dusty and mottled, by the effect of the earth and the ash that it had carried away.

Modern geologists recognize this as a description of a large eruptive plume. Moreover, volcanic activities of the type of Vesuvius in 79 CE are called "Plinian", in recognition of the relevance of the account of Pliny the Younger.

Numerous descriptions and analyses of volcanic eruptions, mainly during the 18th, 19th and early 20th centuries, have served, and still serve, to study the eruptive processes and behavior of volcanoes. Among the most emblematic, we can cite:

– the eruption of Laki (or Lakagigar) (1883) for its climatic effects;

– the Tambora eruption (1815) for its climatic effects and its devastating toll (92,000 victims);

– Krakatau (1883) for its impact on the environment (tsunami, pressure wave, etc.) and its 36,000 victims;

– that of Mount Pelée (1902) which, apart from its 30,000 victims, allowed Alfred Lacroix to document phenomena such as the growth of domes and the flow of pyroclastics;

– more recently, the eruption of Paricutin (1943) where precursors were observed.

4.1.1. *Maps and charts*

In accounts from the 18th and 19th centuries, some explorers developed maps and diagrams that allow us to assess the changes that occurred at those times and since. Emblematic examples of these contributions are found in Hamilton's observation of Vesuvius (1776, illustrated by Fabris) (see Figure 4.1), Bory de Saint-Vincent's Piton de la Fournaise (1804) (see Figures 4.2 and 4.3) or Ellis's Kilauea (1825) (see Figure 4.4).

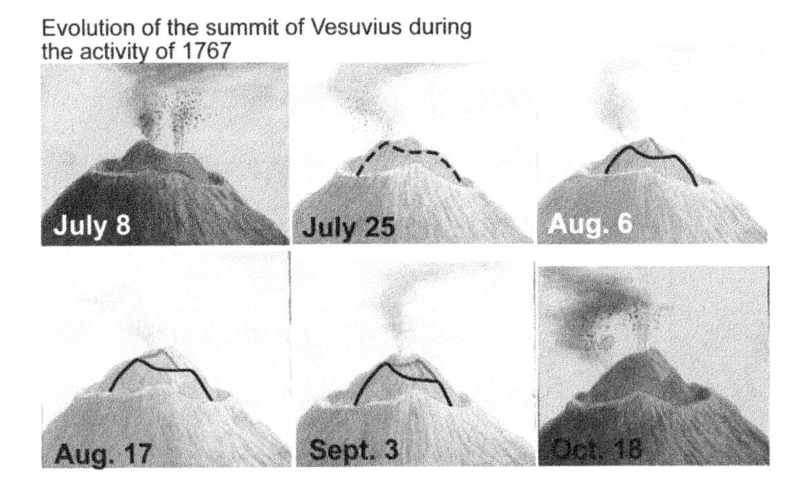

Figure 4.1. *Evolution of the summit of Vesuvius by Hamilton (1776) (drawings by Pietro Fabris). The observations were made from a villa located between Pompeii and Herculaneum. For a color version of this figure, see www.iste.co.uk/lenat/hazards.zip*

Figure 4.2. *Evolution of the summit of Piton de la Fournaise at the end of the 18th century by Bory de Saint-Vincent (1804) from old drawings and his own observations. The blue line represents the outline in 1801. For a color version of this figure, see www.iste.co.uk/lenat/hazards.zip*

Figure 4.3. *A lava lake observed at the top of Piton de la Fournaise by Bory de Saint-Vincent in 1801*

COMMENT ON FIGURE 4.3.– *He gives a precise description of its activity: "The mouth of the volcano presented a vast cauldron of more than two hundred toises in diameter (about 400 m), full, up to the edge, of molten matter, fluid enough to form waves. In the center and where I had seen the sheaves of the 6 brumaire (October 28), rose in dome and fell on themselves, streams of liquid and incandescent lava, while all the surface of the boiler was a little tarnished by a very thin layer of slag. During the day this layer appeared blackish, but at night it let in a certain light. Zig-zag cracks, similar to the tiles of lightning, very numerous, and more or less arranged like the rays of a circumference, gouged the slaggy crust which followed a sort of concentric undulating movement."*

Figure 4.4. *The Kilauea caldera as observed by Ellis in 1823. It was progressively filled by lava until its further collapse in 2018. The collapse/fill cycle of this caldera is well described by Swanson et al. (2012) (drawing from Ellis (1826))*

This evolution leads Bory de Saint-Vincent to propose that cycles of construction and collapse of volcanoes follow one another: "The ridge of volcanoes lowers and rises in turn: we see this at Vesuvius, which is more within reach of observers, and there is every reason to believe that the top of Bourbon Mountain tends towards the same revolution. If it takes place in the centuries to come, the craters that I visited will be encrusted by lava that will raise the mountain to the future subsidence. In the walls of the crater at that time one will observe events which one will not be able to account for, because one will not have known all the states through which the mountain will have passed; and how many times will new forms have known all the states through which the mountain will have passed; and how many times will new forms have been swallowed up by new forms!"

Box 4.1. *Collapses explained by Bory de Saint-Vincent*

4.1.2. *Quantitative data and insights into volcanic mechanisms*

The first explorers also often tried to quantify their observations, with the means of their time. We can quote the first evaluations of the volume of flows at Piton de la Fournaise (Bory de Saint-Vincent 1804). For example, for the flow of 1774:

> Because of all the circuits that this flow followed, we can give it three thousand toises in length (about 5,800 m), by six hundred in average width (about 1,200 m); by granting it only two and a half toises in depth (about 5 m), we will have four million five hundred thousand cubic toises (about 33 million of m^3) of lava spewed in 1774 by the volcano.

The same Bory de Saint-Vincent puts forward the hypothesis that the prismation of the lava flows results from a thermal withdrawal during the cooling and he discusses the formation of the craters and calderas of the Piton de la Fournaise by collapse mechanisms. He notes that "these events, moreover, arrive only after prodigious eruptions" and that the Dolomieu crater such as he observed it in 1801 owed its formation to a collapse ("undoubtedly, the lava, rejected this year by the mountain, had left in its dome some large cavities whose vaults collapsed on themselves").

4.2. The development of instrumental surveillance: late 19th–early 20th centuries to 1970s

4.2.1. *Volcanic observatories*

4.2.1.1. *The Vesuvius Observatory*

The Vesuvius Observatory (*Osservatorio Vesuviano*) was the first volcanological observatory created. This precociousness is linked to the conjunction of a sustained volcanic activity in an inhabited area and the cultural and scientific vivacity of Naples (and Europe in general) at the beginning of the 19th century. In fact, since its devastating eruption in 1631, Vesuvius was in almost continuous activity until 1944. This activity was widely illustrated by the famous gouaches of Vesuvius, then by photographs from the late 19th century. The observatory was built on the slopes of Vesuvius from 1841 to 1848 (see Figure 4.5), under Ferdinand II (king of the Two Sicilies), great promoter of scientific, technical and cultural progress. It was entrusted to Macedonio Melloni, a renowned scientist in the fields of thermal radiation and magnetism. The observatory was initially dedicated to meteorology and magnetism. Melloni's reign was brief, since he was dismissed in 1849 for political reasons. The observatory was revived in 1855 with the appointment of Luigi Palmieri as director (until his death in 1896). In fact, he used the observatory since 1852 for his research. He founded a scientific journal, *The Annals of the Vesuvius Observatory*, in which he gave detailed descriptions of the eruptions from 1855 to 1872. Palmieri designed the first electromagnetic seismograph (see Figure 4.6). It was not actually a seismograph in the present sense (recording waves) but a set of *seismoscopes* connected to a clockwork, capable of detecting and recording the onset of vertical and horizontal movements and the duration of the signals (Dewey and Byerly 1969). Nevertheless, this was the world's *first instrumental monitoring of a volcano*. A portable version, carried in boxes by mules, was even used for Vesuvius (Palmieri 1874).

Palmieri had a very advanced vision of what should be implemented for the monitoring of Vesuvius. Thus, in a communication to the Pontanian Academy (*Accademia Pontaniana*) of Naples, he outlined four points to be developed (Borgstrom et al. 1999):

– recording of the temperature, volume and composition of the products emitted by the fumaroles on and around Vesuvius;

– daily recording of thermal gradients at various locations;

– recording of seismic and geomagnetic signals and of atmospheric and telluric electrical activity;

– a comparison of these signals with those from Rome and Naples in order to separate regional variations from variations of volcanic origin.

Figure 4.5. *The observatory and the cone of Vesuvius represented on the cover of the first issue of the Annals of the Observatory in 1859*

a) b)

Figure 4.6. *Palmieri's "seismoscope" (photo by Giudicepietro et al. (2010)). a) The clocking and recording system. b) The vertical and horizontal motion sensors*

Another famous director was Guiseppe Mercalli from 1911 to 1914 (date of his accidental death in a fire at the observatory). Besides his work in volcanology, Mercalli is also famous for his work in seismology and his seismic scale based on earthquake damage. The last director before the modern period was Giuseppe Imbò, from 1935 to 1970 who, among other things, predicted the eruption of 1944.

It should also be noted that vertical deformation monitoring was initiated very early in this region. Palmieri had tested this method to monitor cracks at Torre del Greco in 1861 (Pingue et al. 2013), and Loperfido installed an 11-km leveling profile between Herculaneum and the summit in 1913 (Pingue et al. 2013). The measurements were repeated in 1919 and, partially, in 1922. Only two benchmarks survived after 1944. In 1905, the Military Geographic Institute installed a precision leveling network in the Phlegraean fields that is still in use (Del Gaudio et al. 2010). Likewise, the study and monitoring of gases was initiated (see Figure 4.7).

Figure 4.7. *Analysis of a fumarole at Vesuvius in 1906 (document observatory of Vesuvius)*

4.2.1.2. *Japanese observatories*

Japan's more than 70 historically active volcanoes represent one of the highest densities of volcanoes in a densely populated area. It is therefore logical that this country has developed very early capabilities for the study and monitoring of volcanic activity. In addition, another major telluric hazard, seismic activity, has also contributed to the development of monitoring techniques.

The pioneer of volcanological monitoring was Fusakichi Omori (1868–1923). He was a disciple of English scientists working at the Imperial University of Tokyo, and in particular of John Milne, who developed, with his colleagues James Ewing and Thomas Gray, a horizontal pendulum seismograph as early as 1880 (Dewey and Byerly 1969). After Milne's departure in 1895, Omori perfected their seismometer, which was then built by the Bosch firm in Germany to become the famous Bosch-Omori seismometer that was used extensively in the early 20th century. This instrument had a signal amplification of about 10 and a natural period of about 20 seconds. It should be noted that before this period, Japan had purchased two Palmieri seismoscopes. They were used in Japan from 1875 to 1883 before being replaced by the seismometers of Milne and his collaborators.

Omori made the first temporary records of volcanic seismic activity in 1910 on Usuzan and Asamayama volcanoes. In 1911, he established the first Japanese volcanological observatory on the flank of Asamayama volcano with the sponsorship of the Imperial Earthquake Investigation Committee and the cooperation of the local meteorological observatory in Nagano. Omori was forced to close the observatory the same winter (due to climatic constraints). The local meteorological observatory of Nagano planned to open, in 1922, a permanent station for meteorological and volcanic observations at the foot of the volcano. Unfortunately, Omori died the following year and the observatory was abandoned.

The second Japanese observatory was installed in 1928 on Asosan volcano under the name of Aso Volcanological Laboratory of Kyoto University. Then, the Seismological Research Institute of the University of Tokyo created the volcanological observatory of Asama volcano. Other volcano observatories established by the Japan Meteorological Agency (JMA) and other institutes and universities would follow.

Deformation monitoring was initiated very early in Japan. A leveling and triangulation network of Japan was established between 1890 and 1900. It was used to describe the deformations during the great eruption of Sakurajima in 1914 (Omori 1916; Yokoyama 1986).

4.2.1.3. *Hawaiian Volcano Observatory*

Hawaiian Volcano Observatory (HVO) is probably the most famous and iconic volcanological observatory. Many publications discuss its founding and development (Apple 1987; Babb et al. 2011; Dvorak 2011; Tilling et al.

2014). It is in this observatory that most of the volcanological monitoring methods have been tested and developed, although we should not forget pioneering research carried out in other countries such as Italy or Japan. This site will allow us to describe the main developments in monitoring until the 1970s.

4.2.1.3.1. The beginnings of HVO

The founder of HVO was Thomas Jaggar (1871–1953). He graduated from Harvard University (PhD in 1897) and was later employed at Harvard University and the Massachusetts Institute of Technology (MIT). His vocation for volcanology was triggered by his visit to Mount Pelée after the catastrophic eruption of May 8, 1902 (he was part of a delegation of American scientists sent there by the U.S. government). In 1906, he went to Naples to study the eruption of Vesuvius, but arrived two days after the end of the eruption. On this occasion, he met Frank Perret. This American engineer, former collaborator of Thomas Edison, was in Naples for health reasons and collaborated with the observatory of Vesuvius. He had improved and built various instruments for the observatory. These impressed Jaggar enormously. He later asked Perret to come to Hawaii to help him with measurements. Jaggar then traveled to other volcanic regions: the Aleutians, Alaska, Japan and Hawaii. The sight of Kilauea and its lava lake in the Halemaumau crater convinced him that, for a variety of reasons, it was a good place to study how volcanoes work. He therefore began to take steps to raise funds to create a permanent observatory. Details of Jaggar's arduous path to raising funds can be found in Dvorak (2011) and Apple (1987).

The fundamental objective of Jaggar was the protection of people and property, based on rigorous scientific knowledge of how volcanoes work. For this purpose, the main missions set for HVO were the following:

– measure and record seismic, eruptive and geodetic phenomena of Hawaii's active volcanoes;

– a geological approach to volcanic deposits and products to understand the evolution of volcanoes; determine the frequency and style of eruptions, in order to assess the hazards;

– systematic sampling and analysis of gases and solids;

– wide communication of the results, both within the scientific community and with the authorities and the public.

Although the official birth of HVO was in 1912, the first work began from July to October 1911. Jaggar could not arrive until January 1912, so he asked E.S. Shepherd (Geophysical Laboratory of the Carnegie Institution of Washington) and Frank A. Perret to make the early measurements and observations. Prior to this, Jaggar had used funds allocated by the Whitney Foundation to MIT to order a Bosch-Omori seismometer in Germany, three Omori seismometers in Japan and to have a Baltimore company make platinum thermometers (the electrical resistance of platinum increases with temperature) capable of being used to measure the temperature in the Halemaumau lava lake. These heavy thermometers, 3 m long, required the use of a cable and a winch mounted on a cart. On July 30, 1911, after two thermometers were lost in the lava lake, a measurement was successful with the last one. It was the *first direct measurement of lava temperature* and the value was 1,000°C. Because of the difficulties of this type of measurement, Shepherd suggested that it would be better to use an optical pyrometer (Jaggar 1917). From July to October 1911, Perret wrote weekly reports (published in a Honolulu newspaper) on the condition of the lava lake.

Figure 4.8. *Omori seismographs inside the HVO "Whitney" vault in 1912 (photo by Thomas Jaggar, USGS/HVO)*

With the arrival of Jaggar in early 1912, construction of a building was undertaken. A history of the successive HVO buildings can be found, for example, in Babb et al. (2011). Seismologist H.O. Wood was responsible for the *installation of the Omori seismometers*, which was completed in August 1912 (Wood 1913) (see Figure 4.8). The most sensitive of the seismometers had an amplification of 120–200.

The two-component Bosch-Omori seismometer arrived from Germany in 1913. This instrument, mechanically and dynamically superior to the others, was to remain the main seismometer of the HVO until 1953 (the Omori seismometers were abandoned in 1923). All seismometers were regularly maintained and improved. HVO even built its own two-component horizontal seismometers, which, around 1928, were even used in other parts of the United States.

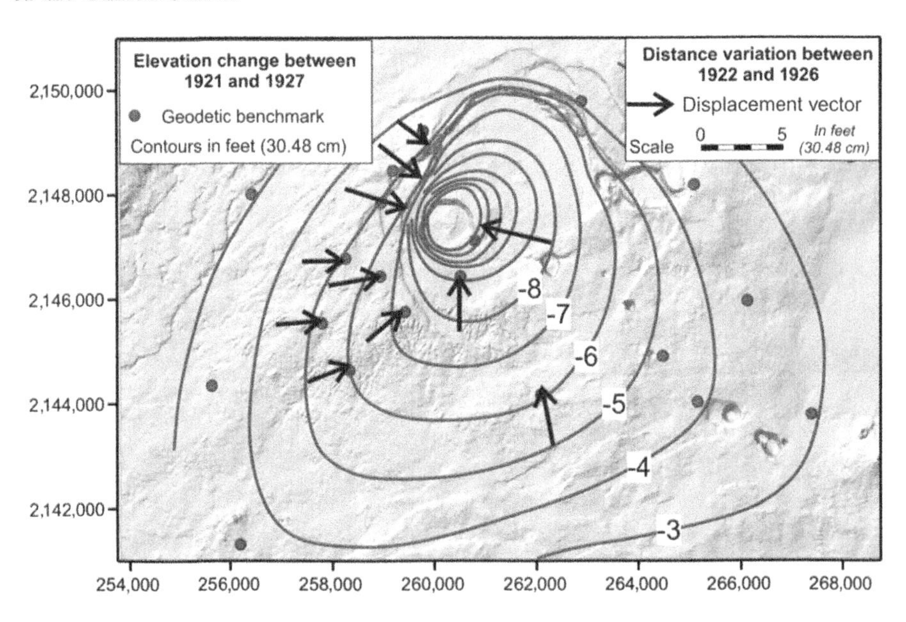

Figure 4.9. *Subsidence contour (up to 4 m) and horizontal displacements (up to 1.6 m) around Halemaumau Crater related to its emptying and collapse in 1924 (after Wilson (1935)). For a color version of this figure, see www.iste.co.uk/lenat/hazards.zip*

Deformation monitoring began very early at Kilauea. A triangulation of the island of Hawaii from 1871 and precision leveling from 1926 formed the basis for this monitoring. A tide gauge installed at Hilo in 1927 could be used as a reference to mean sea level. The work of Wilson (1927, 1935) (see Figure 4.9), comparing measurements framing the collapse of the lava lake at Halemaumau Crater in 1924, demonstrated the potential of deformation measurements in monitoring and analyzing volcanic activity. Previously, measurements with a theodolite had been made to monitor the level of the lava lake. Note that one of the geodetic markers of 1921 in the caldera (SPIT) still exists.

The measurement of *ground tilt* began accidentally at Kilauea. Seismologist Wood noted a drift in the tilt of the Bosch-Omori seismometers after their installation. These data were recorded and compared with temperature, rainfall and lava lake levels. It became clear that most of the signal was related to the deformation of the volcano. *Inclinometry as a method of volcano monitoring was born.* An article by Finch and Jaggar (1929) describes the processing of these data. The sensitivity of these instruments could reach the microradian or less. These measurements were made daily from 1913 to 1963.

Figure 4.10. *Early gas sampling from an active vent in Halemaumau in 1912 (after Day and Shepherd (1913))*

The study of *magmatic gases* was undertaken early in the HVO, thanks to the opportunity provided by the presence of an accessible lava lake. The history of early studies is well described by Sutton and Elias (2014). A.L. Day and E.S. Shepherd of the Carnegie Institution Geophysical Laboratory were the first to take gas samples directly from the lava lake (see Figure 4.10), using tube, bottle and pump systems (Day and Shepherd 1913). Analysis was done at Oahu College and the Carnegie Institution.

4.2.1.3.2. HVO developments until the 1960s

Seismic monitoring

The evolution of HVO seismic monitoring is representative of the evolution of this discipline. Of course, the synergy between the development of regional and global seismic networks and those dedicated to volcanological monitoring should be noted.

HVO seismic monitoring evolved slowly until the arrival of Jerry Eaton in 1953. The history of the networks can be found in Klein and Koyanagi (1980) and Okubo et al. (2014). The early instruments were gradually improved and new ones were even built. But the main difficulties remained having a common and accurate *time base* for different stations and the limited capabilities to *locate and characterize earthquakes*, despite commendable efforts (Jones 1935). Nevertheless, a catalog of more than 17,000 earthquakes exists for the period between 1912 and 1953 (Klein and Wright 2000).

Through his training in seismology, Eaton was able to establish the characteristics of an electromagnetic seismometer suitable for recording the small earthquakes common in Hawaii. The first instruments built under his instructions, named HVO-1 and 2, were short-period vertical seismometers. HVO-2 had an amplification of 40,000 at the 0.2-second frequency. Signals from the field stations were transmitted by telephone cables run in the field. These more sensitive seismometers increased the number of earthquakes recorded at Kilauea by several orders of magnitude (Eaton 1996). However, cable transmission limited their deployment to a small area (a few kilometers) around HVO. Eaton also installed a Wood-Anderson seismometer in 1957 because this instrument allowed the *magnitude* of earthquakes to be determined according to the Richter scale (Richter 1935).

The location of earthquakes was done graphically with travel time tables (Macdonald and Eaton 1964) and the events were classified into families according to their signatures. These advances allowed Eaton and Murata to write their famous paper "How Volcanoes Grow" (1960) in which they show a seismic section of the island of Hawaii including a flexure of the lithosphere. Eaton's 1962 paper (Eaton 1962) clarifies the structure and functioning of the volcano in a remarkable way.

Deformation monitoring

– *Tilt variations*: As mentioned above, Jaggar and his collaborators had understood early on that the drifts of the Omori seismometers were related to the deformation of the volcano and had recorded these tilt variations since 1913. Simple inclinometers were built in the 1920s at HVO, but their signals were not as good as expected. Following a major deflation of the summit during the 1955 eruption, Eaton began the development of better instruments. To avoid the many problems inherent in horizontal pendulums, Eaton chose to develop liquid inclinometers (Eaton 1959) based on earlier work (Michelson 1914; Michelson and Gale 1919) and in particular on work on Japanese volcanoes (Hagiwara 1947; Hagiwara et al. 1951). These instruments consisted of two containers (made from shell casings) containing water, connected by a tube (the water level in each container was measured by a micrometer system). Another tube connected the top of the containers to ensure equal pressure between them (see Figure 4.11). One station was composed of two perpendicular systems in order to resolve the tilt vector. Two short-base stations (about 5 m) were installed in vaults (Whitney and Uwekahuna). The Uwekahuna station is still in use and is a reference for the deformation history of Kilauea. With records made on seismometers as early as 1913, HVO now has more than a century of continuous records of tilt variations (see Figure 4.12). The accuracy of the short-base stations is in the microradian range or less. At the same time, portable water stations were developed for the field (there were a half-dozen of these in 1959), to periodically determine the 2D geometry of tilt variations. These stations had a base of about 50 m. The vessels were placed on a brass benchmark attached to a concrete block (see Figure 4.11(c)). To minimize thermomechanical effects (temperature and sunlight), measurements were made at the end of the night. Accuracy was as high as 0.2 microradians, but experience shows that it was only a few microradians (Decker et al. 2008).

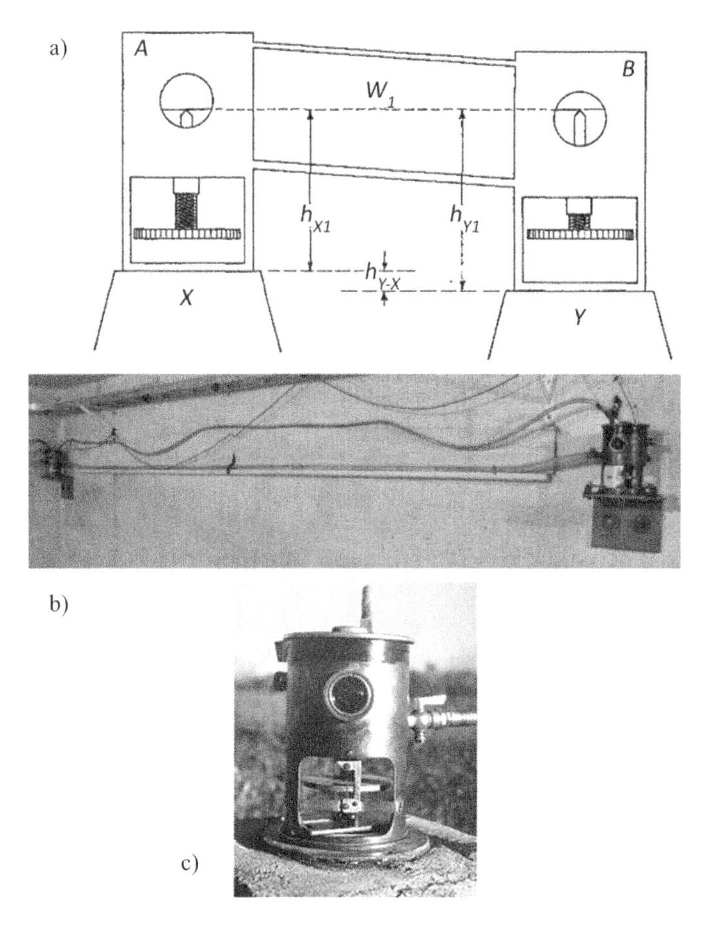

Figure 4.11. *The HVO water inclinometers. a) Principle of leveling between two containers. b) One of the components of the Uwekahuna vault water inclinometer. c) Detail of a container for field measurements (USGS photos)*

This type of leveling for tilt stations was commonly referred to as *wet-tilt*. It will be replaced by *dry-tilt*. The dry-tilt, inspired by the work of E. Tryggvason in Iceland in the early 1960s, replaced hydraulic leveling by optical leveling with a precision level and invar rods. The accuracy was comparable to that of wet-tilt, but the measurements were less demanding and faster (Kinoshita et al. 1974; Yamashita 1981; Decker et al. 2008). The dry-tilt method will be used on many volcanoes.

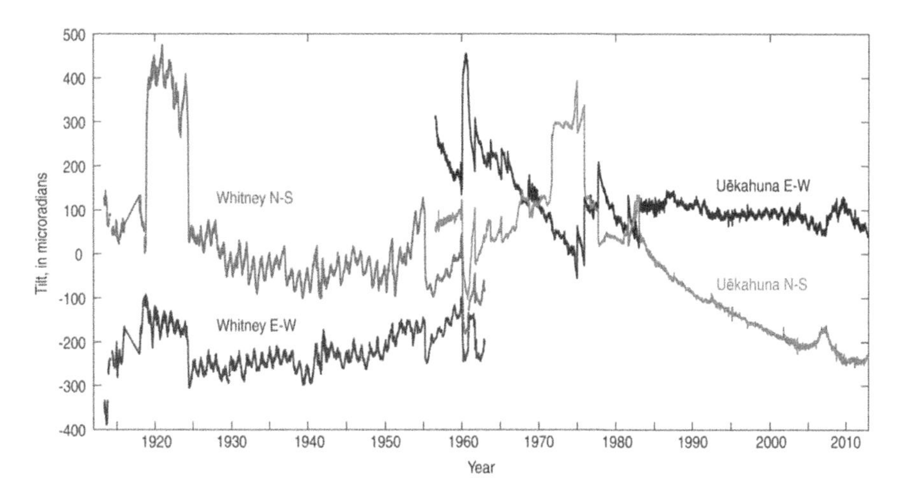

Figure 4.12. *Record of tilt variations at Kilauea from 1913 to the present. For a color version of this figure, see www.iste.co.uk/lenat/hazards.zip*

COMMENT ON FIGURE 4.12.– *The first part is based on the seismometers of the Whitney vault and the second part on the water inclinometer of the Uwekahuna vault. The offset between the curves comes from the difference in location between the two vaults.*

A short-base (1 m) mercury inclinometer based on measuring the capacitance between the liquid and a metal plate (separated from the mercury by air) would be installed in parallel with the water inclinometer in the Uwekahuna vault in 1966 and would allow continuous monitoring of inclination at Kilauea (Dzurisin 2007). This paved the way for permanent inclinometers to be installed in small boreholes beginning in the 1980s.

– *Measurement of horizontal displacements by distance meter*: HVO was a pioneer in the use of electronic distance meters (or geodimeters). The first measurements were made in 1964 across the Kilauea caldera (about 3 km) (Decker et al. 1966), then in 1965 across the Mauna Loa caldera (Decker and Wright 1968). This method, known by the acronym EDM (Electronic Distance Measurement), also allows triangulations to be made more easily and quickly than with theodolites. The EDM network of HVO has grown to include more than 750 measurement lines. Today, Global Navigation Satellite System (GNSS) methods have replaced EDM, after an overlap between the two types of measurement in the years 1990–2000 (see Figure 4.13).

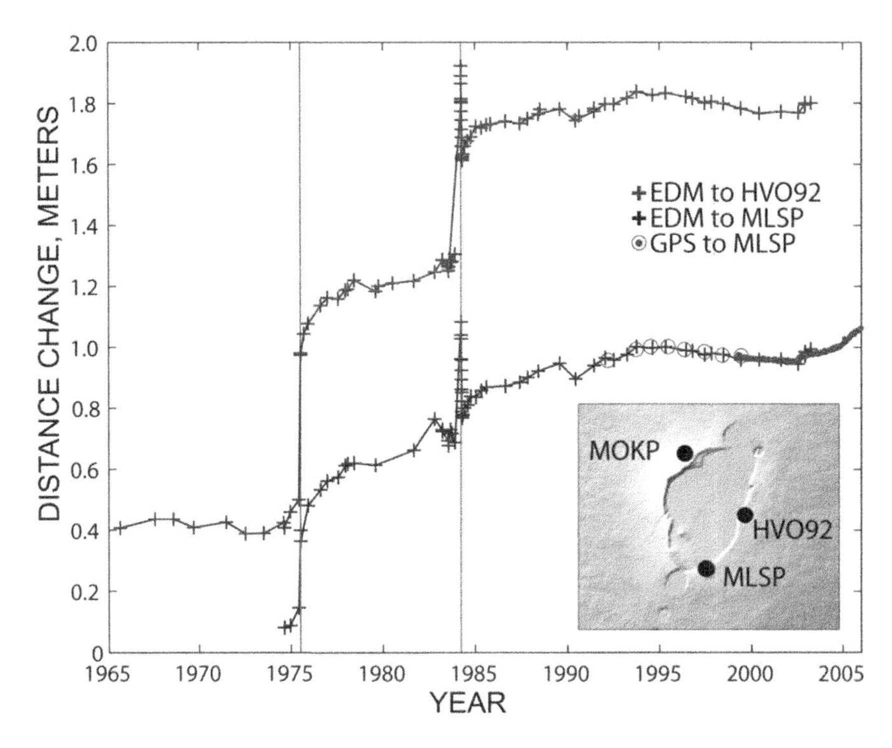

Figure 4.13. *EDM measurements across the caldera of the volcano Mauna Loa since 1965 and overlap with GPS measurements since the late 1990s (USGS document). For a color version of this figure, see www.iste.co.uk/lenat/hazards.zip*

– *Tide gauge*: the tide gauge installed at Hilo in 1927 and the Honolulu tide gauge installed in 1946 have been used (Moore 1987; Caccamise II et al. 2005) to identify and quantify subsidence on the island of Hawaii.

Gas monitoring

Shepherd and Jaggar published analyses of gases collected in Halemaumau between 1912 and 1919 (Shepherd 1925; Jaggar 1940). After the lava lake disappeared in 1924, it was not until 1959 that further magmatic gas studies were conducted. In the meantime, studies were conducted on fumarole gases at Kilauea (Sulfur Banks) (Ballard 1940; Jaggar 1940). Nevertheless, the analyses of Shepherd and Jaggar were later used to refine their equilibrium condition (Matsuo 1962; Nordlie 1971). Around the 1960s, there was a new interest in volcanic gases in Japan and the USSR (Naboko 1959; Matsuo 1960; Iwasaki et al. 1962). Gas samples were collected from eruptions at Kilauea in

1959 and 1960 (Heald et al. 1963). In 1968, Naughton et al. (1969) made what is probably the first measurement of the gas composition of a plume by *infrared remote sensing*. Another type of remote sensing measurement was introduced at Kilauea in 1975 by Stoiber and Malone (1975), the measurement of SO_2 by UV correlation spectrometry (COSPEC). This measurement became regular from 1979, so a series spanning four decades is available at Kilauea (see Figure 4.14).

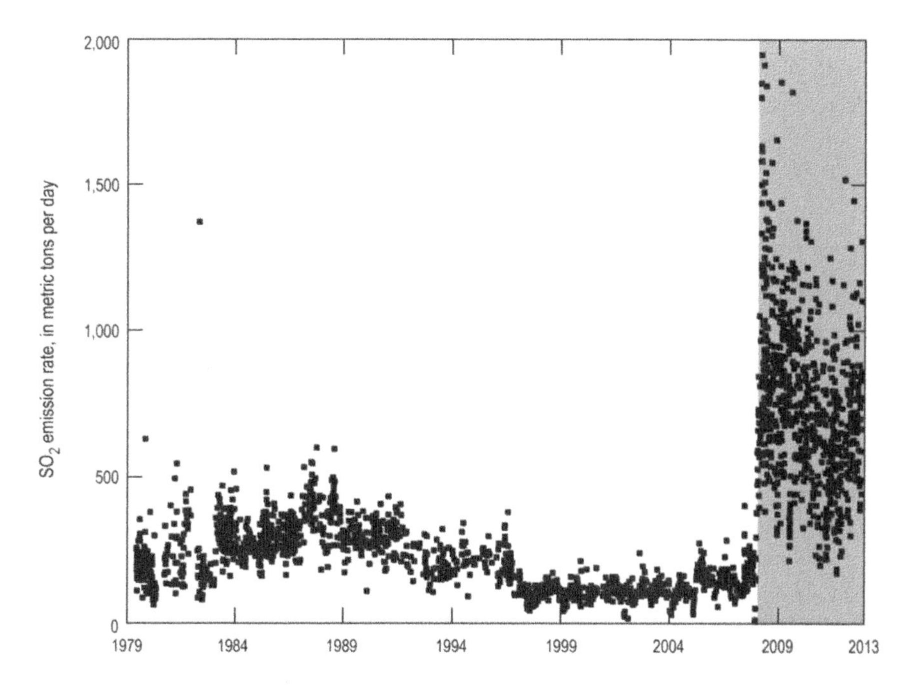

Figure 4.14. *SO_2 emissions from the Kilauea caldera since 1979. The increase from 2008 is associated with the formation of a new lava lake in Halemaumau Crater (from Sutton and Elias (2014))*

Monitoring of products

A systematic follow-up of the petrology and geochemistry of the products of the eruptions has been carried out since 1952.

4.2.1.4. *French volcanological observatories*

The first volcanological observatory built in France was in Martinique in 1903, after the catastrophic eruption of 1902 and following the intervention of Alfred Lacroix (see also Chapter 1). This observatory operated until 1925. As the volcano no longer showed any signs of activity, the observatory was closed only for 4 years before the eruption of 1929–1932. The volcanological and seismological observatory of Martinique succeeded it in 1935 and that of Guadeloupe was created in 1950. These West Indian observatories monitor, of course, the volcanic activity of the volcanoes on which they are installed, but they also operate a regional seismic network to monitor tectonic seismic activity and tsunamis at the scale of the Caribbean arc. The volcanological observatory of Piton de la Fournaise was created at the end of 1979.

4.2.2. *The modern period: impact of digital and space*

It is impossible to date precisely the beginning of a modern period of volcanological monitoring because there is always a continuum in the developments of science. But we can certainly say that from the 1970s onward there was a conjunction between important volcanic crises and the rise of digital and space techniques. All this contributed to an acceleration of innovations in the methods and capacities of monitoring. At the same time, and this is inseparable, fundamental research on volcanic processes has followed the same path. As an anecdote, it probably seems strange to today's young researchers that serious and innovative science could have been done without computers and GPS. We must therefore pay tribute to all the excellent scientists who preceded the current period and advanced science with less sophisticated tools but with a lot of thought and technical tricks. How will the researchers of the end of the 21st century view the current era?

4.2.2.1. *Significant eruptions in the 1970s*

In 1973, the eruption of the Eldfell volcano (Thorarinsson et al. 1973) on the island of Heimaey in southern Iceland led to the evacuation of all 5,300 inhabitants and lava flows threatened to invade its economic heartland, the harbor. The island had no means of surveillance and the eruption surprised the inhabitants in the middle of the night. Only the major precursor earthquakes had been detected by relatively remote stations in southern Iceland, without the possibility of locating them precisely.

In 1976, the Soufrière disaster in Guadeloupe (Feuillard et al. 1983) was accompanied by earthquakes, phreatic explosions, gas emissions, landslides and a mudflow. A total of 70,000 people were evacuated for several months. The crisis revealed the deficiencies in the knowledge of the volcano, in the monitoring networks, in the organization and skills of the French scientific community and in communication in general.

On January 10, 1977, at about 10:00 a.m., fractures opened on the flanks of Nyiragongo volcano (Tazieff 1977; Durieux 2002; Komorowski et al. 2002). The crater was occupied by a lava lake that was instantly drained by the fractures to the north and south of the cone. The altitude difference of about 1,000 m between the top of the lava lake and the fractures explains the exceptional flow of lava: about 20 million m^3 of lava were emitted in less than an hour. The combination of steep slopes, the flow rate and the very low viscosity of the lava allowed for rarely observed flow speeds (up to 60 km/h). Several hundred people were killed because they could not escape from the flow. There were no monitoring means in this region. A seismic station located about 100 km away (Hamaguchi et al. 1992) recorded sustained seismic activity in the area since December 1976, but these events could have occurred both under the nearby volcano Nyamuragira (erupting from December 23, 1976) and under Nyiragongo or could have been related to East African Rift activity. The inhabitants around the volcano had also felt numerous earthquakes before the eruption.

Other major crises could be added, such as the bradyseismic episode of the Phlegraean Fields (Italy) from 1969 to 1974 (Corrado et al. 1976), the eruption of Fuego (Guatemala) in 1974 or of the Soufrière St. Vincent in 1971–1972.

Probably the most prominent eruption of this period was the May 1980 eruption of Mount St. Helens. Apart from the eruptions on the Alaska Peninsula (including the famous Novarupta eruption in 1912 (Hildreth and Fierstein 2012)), the last eruptions in North America were the Lassen Peak eruptions between 1914 and 1917 (Diller 1916). Thus, there was no modern experience in monitoring and managing a volcanic crisis on the continent. However, the United States had developed research on volcanic systems and considerable expertise on monitoring at HVO. The awakening of the volcano in March 1980 was taken very seriously because of the work of Dwight Crandell, Donald Mullineaux and colleagues (Crandell et al. 1975; Crandell and Mullineaux 1978) that demonstrated the potentially dangerous activity of this volcano (see also Chapter 1 on this topic). The combination of geological knowledge and monitoring expertise would allow the crisis to be monitored with a precision never before achieved on an explosive volcano (Lipman and Mullineaux 1981). Instrumental monitoring networks were quickly set up. A state of emergency was declared on April 3 and a no-go zone was established around the volcano. Although the cataclysmic eruption of May 18 caused considerable damage, the number of victims was reduced to 57, whereas it could have reached thousands if protective measures had not been taken.

It was therefore clear that the impact of volcanic activity could be reduced if a long-term scientific investment was made. This was clearly the lesson that could be learned from the failures of the 1970s and the successful management of the Mount St. Helens crisis. Many countries exposed to volcanic hazards have taken the path of modern volcanology, based on both scientific research and the development of monitoring tools. Several countries concerned by volcanic activity had also undertaken to develop a monitoring of their volcanoes. Among the most important are Japan, Italy, New Zealand, Iceland and Indonesia. In all these countries, instrumental monitoring was just beginning, with varying degrees of quality and density.

In Japan (more than 70 volcanoes with historical activity), in response to a very large ash emission from the Sakurajima volcano in 1974, a national program for the development of volcanology was undertaken. It included a large component of modernization of instrumental monitoring (Shimozuru 1983). By the end of the 1970s, there were 17 volcanological observatories. The monitoring of the 1977 Usu crisis was a good example of the level to which Japan had reached at that time. Data on seismicity, deformation (leveling, EDM, inclinometry and stereophotogrammetry), gravity, temperature magnetism (ground and aerial IR imagery), product distribution (ash and flows) and product composition were acquired.

When the bradyseismic crisis of the Phlegraean Fields in Italy occurred between 1968 and 1972, the monitoring resources were very limited (Corrado et al. 1976). No seismic network had existed before 1970. A network of five telemetered seismic stations was installed between 1970 and 1972. A leveling network installed in 1905 allowed monitoring of vertical deformation, and in 1970 networks of four inclinometers and four tide gauges were installed. The situation was better around Vesuvius with a network of seven seismic stations (Giudicepietro et al. 2010). Networks of tide gauges (4), dry-tilt (4) and leveling (20 km) had been established in 1972. The leveling benchmarks were also used for gravimetric monitoring. A trilateration network of 16 bases was installed in 1975 (Bonasia and Pingue 1981; Pingue et al. 1998, 2000). Further south, at Etna and the Aeolian Islands, the late 1970s and early 1980s were the period during which monitoring networks were progressively established (Villari 1983; Falsaperla et al. 1989).

4.2.2.2. *Development of international communication*

Under the direction of France, a new commission of the IAVCEI was created in 1981: the World Organization of Volcano Observatories[1]. It gathers all the institutes involved in volcanic monitoring (mainly observatories) and aims for communication and cooperation among all its members.

With the emergence of the Internet, the means of communication in the international community have become easier, faster and tighter. One of the pioneering applications was the creation in 1984 of the Volcano listserv by John Fink of the University of Arizona at that time. It was a forum maintained by subscribers (scientists) to exchange information (volcanic crises, conference announcements, thesis offers, etc.) quickly by e-mail. This listserv is still very active and is a collaborative project between the Universities of Arizona and Portland, the Global Volcanism Program (GVP) of the Smithsonian Institution's National Museum of Natural History and IAVCEI. The Global Volcanism Program, begun in 1968, has become a site dedicated to volcanic activity during the Holocene. As such, in addition to providing a catalog of all known Holocene volcanoes, it publishes weekly (Smithsonian-USGS Weekly Volcanic Activity Report) and monthly (Bulletin of the Global Volcanism Network) reports on current volcanic activity, contributed by observatories and various volcanology laboratories around the world. These reports were published in paper form and then, in the early 2000s, were only available online in digital form. This was an improvement over the previous situation where reports of volcanic activity were published by IAVCEI in the form of annual reports of the world volcanic eruptions, which were often delayed in time. Similarly, the Holocene Activity Database has largely replaced the Catalog of Active Volcanoes of the World (CAVW) and other lesser known compilations. CAVW was a 22-volume (by region) IAVCEI publication. The GVP database resulted in the publication of three editions (1981, 1994 and 2010) of *Volcanoes of the World* (Siebert et al. 2010).

Finally, the publication of *Encyclopedia of Volcanoes*, with contributions from about 100 specialists, in 1999 and again in 2015 (Sigurdsson et al.

---

1 https://wovo.iavceivolcano.org/.

2015) marked a milestone on the state of knowledge in the various fields of modern volcanology.

4.2.2.3. *Major recent developments in volcanological monitoring*

We can illustrate the revolution that occurred from the 1970s to the 1980s in volcanological monitoring from developments at HVO, while highlighting others by other institutes or scientists. The new tools and methods are largely described in the other chapters of this book. We will therefore limit ourselves to giving only the essential characteristics.

4.2.2.3.1. Seismic monitoring

Considerable progress has been made in terms of sensors, data transmission, recording and processing.

From 1967 to 1997, HVO used a device (Develocorder) capable of recording 20 seismic traces on microfilm. The arrival times were determined graphically on a projection of the microfilm and the possibility of zooming improved the accuracy of reading the times to about 0.05 s (compared to 0.1 s on paper records on drums). Since 1970, the location of earthquakes has been calculated by computer, with a program considering first a medium composed of horizontal layers with constant velocity and then, from 1979, layers with a velocity gradient, taking into account the altitude of the stations and a calculation of the magnitude (HYPOINVERSE program of Klein (1978), which will be regularly improved and developed). In parallel with the Develocorder system, an analog magnetic recorder will allow, from 1977 to 1992, the recording of 112 seismic traces (see Figure 4.15). The next step was the digital acquisition of the signals via an analog/digital converter. At HVO, a system called CUSP (Caltech-USGS Seismic Processing) was installed in 1985. The advantage of this type of development was that it allowed near-real-time automation of processes such as hypocenter determination (via automatic detection of wave arrival times at the various stations). With the advent of personal computers in the 1980s, programs were developed for these computers for both the acquisition and processing of seismic data. One of the most prominent is the Seismological Software Library distributed by the International Association of Seismology and Physics of the Earth's Interior (IASPEI) (Lee 1989). The second generation of such libraries would come from the 1990s with the Earthworm package (Johnson et al. 1995) and then SeisComP3 in the 2000s (Hanka et al. 2000; Olivieri and Clinton 2012). These software packages are open source and modular, allowing the community to develop or adapt modules for different functions.

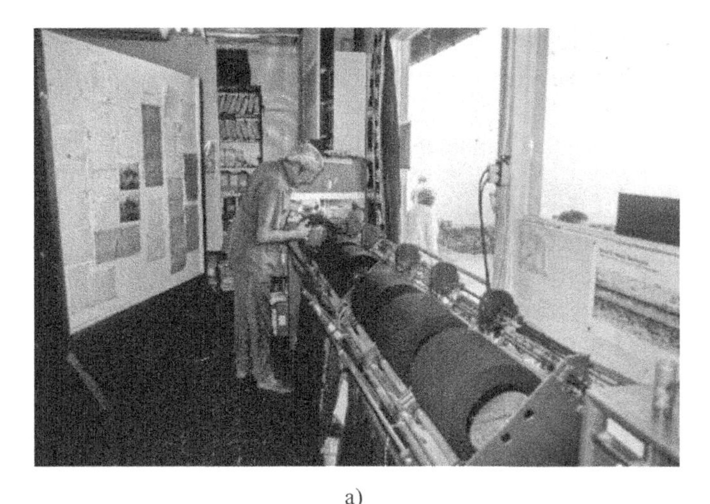

a)

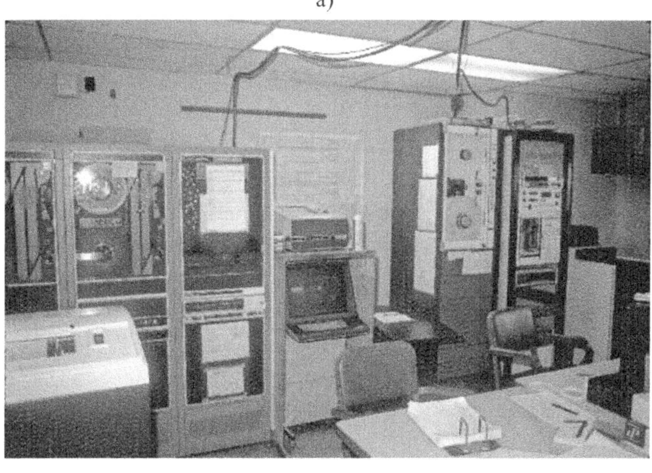

b)

Figure 4.15. *a) Seismic recording drums at HVO in 1980. Before the digital era, these graphic recordings allowed researchers to visualize a day of seismicity on selected stations. b) Recording room (tapes) and digital processing (computer) at HVO in 1980 (photos J.-F. Lénat)*

In the 1970s, seismic monitoring was mainly based on the use of short-period seismometers (~0.1–1 s), whereas seismic signals cover a much wider range, forming a continuum with what is referred to as deformation. The advent of broadband seismometers (~0.1–100 s), high-dynamic digital recordings and powerful analysis techniques has marked the development of new fields of investigation, particularly for signals associated with magma and fluid movements (Neuberg et al. 1994; Chouet 2003; Chouet and Matoza 2013).

The transmission of signals (seismic or other) has also undergone significant changes. Radio links have evolved from the transmission of analog signals to the transmission of digital signals with various protocols (e.g. in Wi-Fi). Part of the transmissions can now be made via the Internet, which allows an evolution toward a global connection of networks. In the early 1970s, an experiment (Endo et al. 1974) was done to transmit seismic and inclinometric data from 15 volcanoes by satellite. The seismic information was compacted into the number of events detected for four levels of amplitude. Similar experiments were done around the 1980s by French researchers using the ARGOS system.

The fields opened by modern volcanic seismology are numerous. They will be described in the chapter devoted to seismic monitoring in Volume 2 of this series of books. Among them, we can mention: automatic detection, localization and analysis of signals in near-real time, seismic interferometry of coda and ambient noise to detect velocity variations in buildings, passive tomography, using ambient noise, monitoring of intrusions, relative localization of earthquakes, identification of active zones in fluid, gas and magma conduits from Very Long-Period (VLP) and Long-Period (LP) signals, detection of external volcanic processes such as explosions, lahars, pyroclastic flows or block falls, etc.

Finally, in related fields, the use of infrasound sensor arrays has developed to monitor and study explosions or other surface phenomena such as pyroclastic flows (Vergniolle et al. 2004; Matoza et al. 2009; Fee and Matoza 2013), and lightning strikes in plumes are recorded and localized by Very High-Frequency (VHF) receivers (Thomas et al. 2007).

4.2.2.3.2. Deformation monitoring

While ground motion monitoring has remained one of the primary methods of monitoring (see Figure 4.16), its practice has evolved significantly since the 1970s. Daniel Dzurisin's (2007) book provides a comprehensive review. A real revolution came with the development of space-based GPS and radar interferometry techniques. On the ground, continuous monitoring techniques using inclinometers or dilatometers have been developed with the increase in technological capacities for recording and transmission. Techniques requiring periodic repetition of field measurements are gradually disappearing as new space-based techniques achieve a comparable degree of accuracy (see Figure 4.13). Numerical deformation modeling techniques were able to develop in parallel with the boom in computing capabilities (Cayol and Cornet 1998). Finally, analog scale experiments for deformation modeling developed for tectonics have been employed in the case of volcanism (Roche et al. 2000).

a)

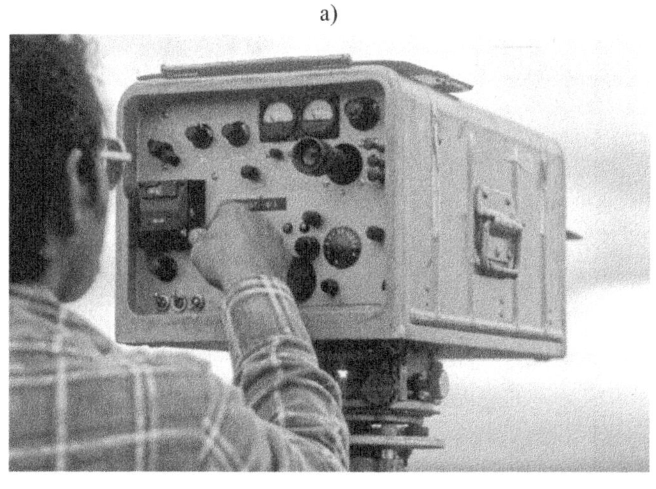

b)

Figure 4.16. *a) Leveling measurements and b) EDM at Kilauea in 1980 (photos J.-F. Lénat)*

4.2.2.3.3. GNSS positioning

Initially, the term GNSS (Global Navigation Satellite System) was confused with GPS because it was the only system available. With time, other systems appeared, the Russian GLONASS, the European Galileo and the Chinese COMPASS (or Beidou), and gradually became operational. After a pre-operational phase between 1978 and 1985, the GPS system (USA) became operational in February 1994.

The use in volcanology has quickly imposed itself not only for localizations at a few meters (sampling, station localization, etc.) with "hiking" type receivers, but especially for the measurement of ground movements (sub-centimetric precision), using sophisticated procedures and receivers. At the beginning, GPS measurements on volcanoes were carried out by periodic reoccupations of landmarks on the ground. Then permanent stations appeared that could be linked to observatories by telemetry.

4.2.2.3.4. Radar interferometry

While all methods of measuring ground deformation provide data at a limited number of points on the surface, the radar interferometry method, which became operational in the early 1990s, was a revolutionary development because of its ability to produce an accurate (sub-centimeter precision) map of displacements over large areas. A seminal paper by Massonnet et al. (1993), applied to earthquake-related displacements in California, marked the beginning of the application of this technique to many problems, including volcanological monitoring (Massonnet and Feigl 1998).

4.2.2.3.5. Continuous deformation records

The need to obtain continuous records of deformation, both to limit field measurements and to record the dynamics of deformation, has been largely met by advances in acquisition and telemetry techniques. This has been seen above for GNSS, but the use of permanent inclinometers and extensometers has also expanded greatly.

The short-base permanent inclinometers developed in this period were either pendulum inclinometers or electrolyte inclinometers. In the first category, we find a magnetoresistance inclinometer developed in Iceland (Sindrason and Olafsson 1978; Tryggvason 1982) and a silica inclinometer developed by Antoine Blum at the IPG in Paris (Blum 1963; Saleh 1986). For electrolyte inclinometers, reference can be made to Westphal et al. (1983) and Powell and Pheifer (2000).

Short-base extensometers (see Figure 4.17) (Jacob et al. 2005; Peltier et al. 2006) and long-base extensometers (Yamazaki et al. 2013) have been installed to monitor fractures or measure strain rate.

Figure 4.17. *A short-base extensometer installed in the early 1980s to monitor a fracture at Piton de la Fournaise (photo J.-L. Cheminée). An invar bar is fixed on one side and a displacement sensor measures the distance variations on the other*

Finally, volume or vector dilatometers have begun to be used for volcanic monitoring (Linde et al. 1993, 2010; Linde and Sacks 2013).

One of the most notable advances has been the development of systems installed in boreholes at a few tens or hundreds of meters depth. At these depths, the sensors are isolated from surface disturbances (thermomechanical effects related to temperature, atmosphere, wind, etc.), and their sensitivity can gain several orders of magnitude (Voight et al. 2010).

4.2.2.3.6. Fluid monitoring

Gas sampling (see Figure 4.18) has benefited from considerable progress with the method developed by Giggenbach (1975). Developments have been made for continuous monitoring of fumarole zones with hydrogen probes (Sato and McGee 1981) or by MultiGAS probes (Shinohara 2005). In the UV range, COSPEC has been replaced by miniaturized DOAS sensors (Weibring et al. 1998; Edmonds et al. 2003). In the IR, FTIR (Mori et al. 1993; Oppenheimer et al. 1998) and FLYSPEC (Horton et al. 2006) instruments have opened up new possibilities. Modern aspects of these methods are described in the chapters on fluid monitoring and remote sensing in Volumes 2 and 3 of this series of books. Finally, the new ease of measurement has allowed the study of diffuse outgassing from volcanoes, whose importance in volcanic processes has been revealed (Chiodini et al. 2001).

Figure 4.18. *Gas sampling on a fumarole zone of Kilauea in 1980 (photo J.-F. Lénat)*

The density (spatial and temporal) of these gas data has opened up new fields for characterizing magmatic transfers and predicting activity (Aiuppa et al. 2007).

4.2.2.3.7. Gravimetry and electrical, electromagnetic and magnetic methods

– *Gravimetric monitoring*: The basic premise of microgravity monitoring is that changes in Earth's gravity can reflect deep mass transfer processes (Dzurisin 2007). However, this signal is very small compared to the average gravity value and is mixed with other types of variations (see Chapter 4, "Gravity Monitoring of Volcanoes", in Volume 3 of this series of books). One of the first experiments on gravity monitoring was that of Lida et al. (1952). Then other measurements were reported from the 1960s to the 1980s (Rose and Stoiber 1969; Eggers and Chavez 1979; Malone 1979; Dzurisin et al. 1980).

Remarkable progress has been made in this method thanks to:

– the improvement of instruments (relative and absolute gravimeters);

– the possibility of making continuous records;

– the greater ease of obtaining precise leveling of stations (GNSS and INSAR space methods);

– digital capabilities for data processing and modeling.

Summaries of the current state of the art are given by Carbone et al. (2017) and Van Camp et al. (2017).

– *Electrical, electromagnetic and magnetic methods*: these methods are best known for their use in the structural study of volcanic edifices. However, the parameters they describe (electrical resistivity and magnetization) can vary over time in response to volcanic activity. Scientists have therefore logically sought to use these methods for monitoring. Pioneering experiments have been conducted, in particular in direct current on Japanese volcanoes (Yukutake et al. 1983, 1987), in EM on Kilauea (Jackson et al. 1985) and in magnetism (Uyeda 1961; Johnston and Stacey 1969; Davis et al. 1973; Pozzi et al. 1979). The application of these methods has developed strongly since then and has led to numerous works.

A special case is the use of the Very-Low-Frequency (VLF) electromagnetic technique used at Kilauea to detect active lava tunnels and estimate their lava flow (Jackson et al. 1988; Kauahikaua et al. 1996, 1998; Sutton et al. 2003).

Finally, the spontaneous polarization (SP) method has been developed extensively over the last few decades to map and track fluid movement in buildings (Lénat 1995).

4.2.2.3.8. The arrival of remote sensing techniques

The acquisition of data on an object without contact with it covers many fields (e.g. we can include seismology), but the term remote sensing is generally reserved for techniques using electromagnetic waves. These techniques took off considerably during the Second World War (radar in particular) and then during the conquest of space since the 1960s. The use of these techniques, on the ground or at altitude, for the monitoring of volcanic activity has developed considerably during the last decades. We are therefore in this field in the contemporary history, and developments are described in the chapters on monitoring deformation, fluids and chapters dedicated to remote sensing in Volume 2 of this series of books.

4.2.2.4. *Disasters, communication and risks*

The impact of volcanic crises on the environment and society has become a major area of research in recent decades. The 1985 Nevado del Ruiz eruption in Colombia triggered an unprecedented reflection on the role of volcanology in society (Sigurdsson and Carey 1986; Tilling 1989; Voight 1990; Williams 1990). From that time on, much work has been done on hazard and risk assessment, risk perception, physical and psychological impacts of volcanic activity on people, environmental impacts, communication with authorities, the press, and the public, and volcanic risk education. This underlines the fact that volcanology is both a fundamental and an applied science.

4.3. Acknowledgments

We thank Franck Donnadieu and Olivier Roche for their meticulous review of this chapter and their pertinent suggestions.

4.4. References

Aiuppa, A., Moretti, R., Federico, C., Giudice, G., Gurrieri, S., Liuzzo, M. et al. (2007). Forecasting Etna eruptions by real-time observation of volcanic gas composition. *Geology*, 35(12), 1115–1118.

Apple, R.A. (1987). Thomas A. Jaggar, Jr., and the Hawaiian Volcano Observatory. In *Volcanism in Hawaii*, Volume 2, Decker, R.W., Wright, T.L., Stauffer, P.H. (eds). U.S. Geological Survey, Reston.

Babb, J.L., Kauahikaua, J.P., Tilling, R.I. (2011). The story of the Hawaiian Volcano Observatory – A remarkable first 100 years of tracking eruptions and earthquakes. *U.S. Geological Survey General Information Product*, 135(60).

Ballard, S.S. (1940). A chemical study of Kilauea solfataric gases; 1938–1940. In *The Volcano Letter*, Fiske, R.S., Simkin, T., Nielsen, E.A. (eds). Smithsonian Institution Press, Washington.

Blum, P.A. (1963). Contribution à l'étude des variations de la verticale en un lieu. *Ann. Géophys.*, 19, 215–243.

Bonasia, V. and Pingue, F. (1981). Ground deformations on Mt. Vesuvius from 1977 to 1981. *Bulletin Volcanologique*, 44(3), 513–520.

Borgstrom, S., De Lucia, M., Nave, R. (1999). Luigi Palmieri: First scientific bases for geophysical surveillance in Mt. Vesuvius area. *Annals of Geophysics*, 42(3), 587–590 [Online]. Available at: https://www.annalsofgeophysics.eu/index.php/annals/article/view/3741 [Accessed March 1, 2022].

Bory de Saint-Vincent, J.B.G.M. (1804). *Voyage dans les quatre principales îles des mers d'Afrique*. Buisson, Paris.

Caccamise II, D.J., Merrifield, M.A., Bevis, M., Foster, J., Firing, Y.L., Schenewerk, M.S. et al. (2005). Sea level rise at Honolulu and Hilo, Hawaii: GPS estimates of differential land motion. *Geophysical Research Letters*, 32(3). doi.org/10.1029/2004GL021380.

Carbone, D., Poland, M.P., Diament, M., Greco, F. (2017). The added value of time-variable microgravimetry to the understanding of how volcanoes work. *Earth-Science Reviews*, 169, 146–179.

Cayol, V. and Cornet, F.H. (1998). Three-dimensional modeling of the 1983–1984 eruption at Piton de la Fournaise Volcano, Réunion Island. *Journal of Geophysical Research: Solid Earth*, 103(B8), 18025–18037.

Chiodini, G., Frondini, F., Cardellini, C., Granieri, D., Marini, L., Ventura, G. (2001). CO_2 degassing and energy release at Solfatara volcano, Campi Flegrei, Italy. *Journal of Geophysical Research*, 106, 16213–16221.

Chouet, B. (1996). New methods and future trends in seismological volcano monitoring. In *Monitoring and Mitigation of Volcano Hazards*, Scarpa, R., Tilling, R.I. (eds). Springer-Verlag, New York.

Chouet, B. (2003). Volcano seismology. *Pure and Applied Geophysics*, 160(3/4), 739–788.

Chouet, B.A. and Matoza, R.S. (2013). A multi-decadal view of seismic methods for detecting precursors of magma movement and eruption. *Journal of Volcanology and Geothermal Research*, 252, 108–175.

Corrado, G.I., Guerra, I., Lo Bascio, A., Luongo, G., Rampoldi, R. (1976). Inflation and microearthquake activity of Phlegraean Fields, Italy. *Bulletin of Volcanology*, 40(3), 169–188.

Crandell, D.R. and Mullineaux, D.R. (1978). Potential hazards from future eruptions of Mount St. Helens volcano, Washington. *U.S. Geological Survey Bulletin*, 1383–C, 26.

Crandell, D.R., Mullineaux, D.R., Meyer, R. (1975). Mount St. Helens volcano; recent and future behavior. *Science*, 187(4175), 438–441.

Davis, P.M., Jackson, D.B., Field, J., Stacey, F.D. (1973). Kilauea Volcano, Hawaii: A search for the volcanomagnetic effect. *Science*, 180(4081), 73–74.

Day, A.L. and Shepherd, E.S. (1913). Water and volcanic activity. *GSA Bulletin*, 24(1), 573–606.

De Boer, J.Z. and Sanders, D.T. (2002). Volcanoes in human history. Princeton University Press, Princeton [Online]. Available at: http://www.jstor.org/stable/j.ctt7s08n.

Decker, R.W. and Wright, T.L. (1968). Deformation measurements on Mauna Loa Volcano, Hawaii. *Bulletin Volcanologique*, 32, 401–402.

Decker, R.W., Hill, D.P., Wright, T.L. (1966). Deformation measurements on Kilauea Volcano, Hawaii. *Bulletin Volcanologique*, 29, 721–731.

Decker, R.W., Okamura, A., Miklius, A., Poland, M.P. (2008). Evolution of deformation studies on active Hawaiian volcanoes. *Scientific Investigations Report*. doi.org/10.3133/sir20085090.

Del Gaudio, C., Aquino, I., Ricciardi, G.P., Ricco, C., Scandone, R. (2010). Unrest episodes at Campi Flegrei: A reconstruction of vertical ground movements during 1905–2009. *Journal of Volcanology and Geothermal Research*, 195(1), 48–56.

Dewey, J. and Byerly, P. (1969). The early history of seismometry (to 1900). *Bulletin of the Seismological Society of America*, 59(1), 183–227.

Diller, J.S. (1916). The volcanic history of Lassen Peak. *Science*, 43, 727–733.

Durieux, J. (2002). Nyiragongo: The January 10th 1977 eruption. *Acta Vulcanologica*, 14–15(1/2), 145–148.

Dvorak, J. (2011). The origin of the Hawaiian Volcano Observatory. *Physics Today*, 64(5), 32–37.

Dzurisin, D. (2007). *Volcano Deformation. New Geodetic Monitoring Techniques*, 1st edition. Springer-Verlag, Berlin.

Dzurisin, D., Anderson, L.A., Eaton, G.P., Koyanagi, R.Y., Lipman, P.W., Lockwood, J.P. et al. (1980). Geophysical observations of Kilauea volcano, Hawaii 2: Constraints on the magma supply during November 1975–September 1977. *Journal of Volcanology and Geothermal Research*, 7(3/4), 241–269.

Eaton, J.P. (1959). A portable water-tube tiltmeter. *Bulletin of the Seismological Society of America*, 49(4), 301–316.

Eaton, J.P. (1962). Crustal structure and volcanism in Hawaii. *The Crust of the Pacific Basin*. doi.org/10.1029/GM006p0013.

Eaton, J.P. (1996). Microearthquake seismology in USGS volcano and earthquake studies; 1953–1995. Open-File Report. doi.org/10.3133/ofr9654.

Eaton, J.P. and Murata, K.J. (1960). How volcanoes grow. *Science*, 132(3432), 925–938.

Edmonds, M., Herd, R.A., Galle, B., Oppenheimer, C.M. (2003). Automated, high time-resolution measurements of SO_2 flux at Soufrière Hills Volcano, Montserrat. *Bulletin of Volcanology*, 65(8), 578–586.

Eggers, A.A. and Chavez, D. (1979). Temporal gravity variations at Pacaya volcano, Guatemala. *Journal of Volcanology and Geothermal Research*, 6(3), 391–402.

Ellis, W. (1826). *Narrative of a Tour through Hawaii, or, Owhyhee; with Observations on the Natural History of the Sandwich Islands, and Remarks on the Manners, Customs, Traditions, History, and Language of their Inhabitants*. Fisher, London.

Endo, E.T., Ward, P.L., Harlow, D.H., Allen, R.V., Eaton, J.P. (1974). A prototype global volcano surveillance system monitoring seismic activity and tilt. *Bulletin Volcanique*, 38(1), 315–344.

Falsaperla, S., Frazzetta, G., Neri, G., Nunnari, G., Velardita, R., Villari, L. (1989). *Volcanic Hazards: Assessment and Monitoring*. doi.org/10.1007/978-3-642-73759-6_20.

Fee, D. and Matoza, R.S. (2013). An overview of volcano infrasound: From Hawaiian to Plinian, local to global. *Journal of Volcanology and Geothermal Research*, 249, 123–139.

Feuillard, M., Allegre, C.J., Brandeis, G., Gaulon, R., Mouel, J.L., Le Mercier, J.C. et al. (1983). The 1975–1977 crisis of la Soufriere de Guadeloupe (F.W.I): A still-born magmatic eruption. *Journal of Volcanology and Geothermal Research*, 16(3), 317–334.

Finch, R.H. and Jaggar, T.A. (1929). Tilt records for thirteen years at the Hawaiian Volcano Observatory 1. *Bulletin of the Seismological Society of America*, 19(1), 38–51.

Fisher, R.V., Heiken, G., Hulen, J.B. (1997). *Volcanoes: Crucibles of Change. Choice Reviews Online*, Volume 35. Princeton University Press, Princeton.

Giggenbach, W.F. (1975). A simple method for the collection and analysis of volcanic gas samples. *Bulletin Volcanologique*, 39(1), 132–145.

Giudicepietro, F., Orazi, M., Scarpato, G., Peluso, R., D'Auria, L., Ricciolino, P. et al. (2010). Seismological monitoring of Mount Vesuvius (Italy): More than a century of observations. *Seismological Research Letters*, 81, 625–634.

Hagiwara, T. (1947). Observations of changes in the inclination of the Earth's surface at Mr. Tsukuba. *Bulletin of the Earthquake Research Institute, The University of Tokyo*, 25, 27–32.

Hagiwara, T., Kasahara, K., Yamada, J., Saito, S. (1951). Observation of the deformation of the Earth's surface at Aburatsubo, Miura Peninsula. *Bulletin of the Earthquake Research Institute, The University of Tokyo*, 29, 455–468.

Hamaguchi, H., Nishimura, T., Zana, N. (1992). Process of the 1977 Nyiragongo eruption inferred from the analysis of long-period earthquakes and volcanic tremors. *Tectonophysics*, 209, 241–254.

Hamilton, W. and Fabris, P. (1776). Campi Phlegraei. Observations on the volcanos of the two Sicilies. As they have been communicated to the Royal Society of London: Observations sur les volcans des deux Siciles. Telles qu'elles ont été communiquées à la Société Royale de Londres [Online]. Available at: https://www.e-rara.ch/zut/content/titleinfo/2883226 [Accessed March 1, 2022].

Hanka, W., Heinloo, A., Jaeckel, K.H. (2000). Networked seismographs: GEOFON real-time data distribution. *ORFEUS Newsletter*, 2(3), 1–24.

Heald, E.F., Naughton, J.J., Barnes Jr., I.L. (1963). The chemistry of volcanic gases 2: Use of equilibrium calculations in the interpretation of volcanic gas samples. *Journal of Geophysical Research (1896–1977)*, 68(2), 545–557.

Hildreth, W. and Fierstein, J. (2012). The Novarupta-Katmai eruption of 1912 – Largest eruption of the twentieth century; centennial perspectives. *U.S. Geological Survey Professional Paper*, 1791, 259 [Online]. Available at: https://pubs.usgs.gov/pp/1791/ [Accessed March 1, 2022].

Horton, K.A., Williams-Jones, G., Garbeil, H., Elias, T., Sutton, A.J., Mouginis-Mark, P. et al. (2006). Real-time measurement of volcanic SO_2 emissions: Validation of a new UV correlation spectrometer (FLYSPEC). *Bulletin of Volcanology*, 68(4), 323–327.

Iida, K., Hayakawa, M., Katayose, K. (1952). Gravity survey of Mihara Volcano, Ooshima Island, and changes in gravity caused by eruption. *Geological Survey of Japan*, 152, 1–28.

Iwasaki, I., Ozawa, T., Yoshida, M., Katsura, T., Iwasaki, B., Kamada, M. (1962). Chemical composition of volcanic gases in Japan. *Bulletin Volcanologique*, 24(1), 23.

Jackson, D.B., Kauahikaua, J., Zablocki, C.J. (1985). Resistivity monitoring of an active volcano using the controlled-source electromagnetic technique: Kilauea volcano, Hawaii. *Journal of Geophysical Research*, 90, 12545–12555.

Jackson, D.B., Kauahikaua, J.P., Hon, K.A., Heliker, C.C. (1988). Rate and variation of magma supply to the active lava lake on the middle east rift zone of Kilauea Volcano, Hawaii. *Geological Society of America Abstracts*, 20(7), A397.

Jacob, T., Beauducel, F., Hammouya, G., David, J., Komorowski, J.-C. (2005). Ten years of extensometry at Soufrière of Guadeloupe: New constraints on the hydrothermal system, Soufriere Hills Volcano. In *Ten Years on International Workshop*, U.W.I. Seismic Research Unit (ed.). University of the West Indies, Tunapuna-Piarco.

Jaggar, T.A. (1917). Report of the Hawaiian Volcano Observatory of the Massachusetts Institute of Technology and the Hawaii Volcano Research Association (dated January–March 1912 on cover). Massachusetts Institute of Technology Society of Art, Cambridge.

Jaggar, T.A. (1940). Magmatic gases. *American Journal of Science*, 238(5), 313–353.

Johnson, C.E., Bittenbinder, A., Bogaert, B.L.D., Kohler, W.K. (1995). Earthworm: A flexible approach to seismic network processing. *IRIS Newsletter*, 14(2), 1–4.

Johnston, M.J.S. and Stacey, F.D. (1969). Transient magnetic anomalies accompanying volcanic eruptions in New Zealand. *Nature*, 224(5226), 1289–1290.

Jones, A.E. (1935). Hawaiian travel times. *Bulletin of the Seismological Society of America*, 25(1), 33–61.

Kauahikaua, J.P., Mangan, M.T., Heliker, C.C., Mattox, T.N. (1996). A quantitative look at the demise of a basaltic vent; the death of Kupaianaha, Kilauea Volcano, Hawai'i. *Bulletin of Volcanology*, 57(8), 641–648.

Kauahikaua, J.P., Cashman, K.V., Mattox, T.N., Heliker, C.C., Hon, K.A., Mangan, M.T., Thornber, C.R. (1998). Observations on basaltic lava streams in tubes from Kilauea Volcano, island of Hawaii. *Journal of Geophysical Research*, 103(B11), 27303–27323.

Kinoshita, W.T., Swanson, D.A., Jackson, D.B. (1974). The measurement of crustal deformation related to volcanic activity at Kilauea Volcano, Hawaii. In *Physical Volcanology*, Volume 6, Civetta, L., Gasparini, P., Luongo, G., Rappola, A. (eds). Elsevier, Amsterdam.

Klein, F.W. (1978). Hypocenter location program, HYPOINVERSE; part I. Users guide to versions 1, 2, 3, and 4; part II. Source listings and notes. *U.S. Geological Survey Open-File Report*, 113, 78–694.

Klein, F.W. and Koyanagi, R.Y. (1980). Hawaiian Volcano Observatory seismic network history, 1950–79. Open-File Report. doi.org/10.3133/ofr80302.

Klein, F.W. and Wright, T.L. (2000). Catalog of Hawaiian earthquakes, 1823–1959. Professional Paper. doi.org/10.3133/pp1623.

Komorowski, J.-C., Tedesco, D., Kasereka, M., Allard, P., Papale, P., Vaselli, O. et al. (2002). The January 2002 flank eruption of Nyiragongo volcano (Democratic Republic of Congo): Chronology, evidence for a tectonic rift trigger, and impact of lava flows on the city of Goma. *Acta Vulcanologica*, 14(1/2), 27–62.

Lee, W.H.K. (1989). *Toolbox for Seismic Data Acquisition, Processing, and Analysis*, Volume 1. IASPEI Software Library, Menlo Park.

Lénat, J.-F. (1995). *Monitoring Active Volcanoes: Strategies, Procedures and Techniques*, McGuire, B., Murray, J., Kilburn, C.R.J. (eds). UCL Press, London.

Linde, A.T. and Sacks, S. (2013). Continuous monitoring of volcanoes with borehole strainmeters. In *Mauna Loa Revealed: Structure, Composition, History, and Hazards*. doi.org/10.1029/GM092p0171.

Linde, A.T., Agustsson, K., Sacks, I.S., Stefansson, R. (1993). Mechanism of the 1991 eruption of Hekla from continuous borehole strain monitoring. *Nature*, 365, 737–740.

Linde, A.T., Sacks, S., Hidayat, D., Voight, B., Clarke, A., Elsworth, E. et al. (2010). Vulcanian explosion at Soufrière Hills Volcano, Montserrat on March 2004 as revealed by strain data. *Geophysical Research Letters*, 37, L00E07.

Lipman, P.W. and Mullineaux, D.R. (eds) (1981). The 1980 eruptions of Mount St. Helens, Washington. *USGS Professional Paper*, 844 [Online]. Available at: http://pubs.er.usgs.gov/publication/pp1250.

Macdonald, G.A. and Eaton, J.P. (1964). Hawaiian volcanoes during 1955. *US Geological Survey Bulletin*, 1171, 170.

Malone, S.D. (1979). Gravity changes accompanying increased heat emission at Mount Baker, Washington. *Journal of Volcanology and Geothermal Research*, 6(3), 241–256.

Massonnet, D. and Feigl, K.L. (1998). Radar interferometry and its application to changes in the Earth's surface. *Reviews of Geophysics*, 36(4), 441–500.

Massonnet, D., Rossi, M., Carmona, C., Adragna, F., Peltzer, G., Feigl, K., Rabaute, T. (1993). The displacement field of the Landers earthquake mapped by radar interferometry. *Nature*, 364, 138–142.

Matoza, R., Fee, D., Garces, M., Seiner, J.M., Ramon, P.A., Hedlin, M.A. (2009). Infrasonic jet noise from volcanic eruptions. *Geophysical Research Letters*, 36, L08303.

Matsuo, S. (1960). On the origin of volcanic gases. *Journal of Earth Sciences, Nagoya University*, 8, 222–245 [Online]. Available at: https://ci.nii.ac.jp/naid/10004204492/en/.

Matsuo, S. (1962). Establishment of chemical equilibrium in the volcanic gas obtained from the lava lake of Kilauea, Hawaii. *Bulletin Volcanologique*, 24(1), 59–71.

Michelson, A.A. (1914). Preliminary results of measurement of the rigidity of the Earth. *Astrophysical Journal*, 39, 105–128.

Michelson, A.A. and Gale, H.G. (1919). The rigidity of the Earth. *The Journal of Geology*, 27(8), 585–601 [Online]. Available at: http://www.jstor.org/stable/30057989.

Moore, J.G. (1987). Subsidence of the Hawaiian Ridge. In *Volcanism in Hawaii: U.S. Geological Survey Professional Paper 1350*, Decker, R.W., Wright, T.L., Stauffer, P.H. (eds). USGS Publications, 85–100 [Online]. Available at: https://pubs.usgs.gov/pp/1987/1350 [Accessed March 1, 2022].

Mori, T., Notsu, K., Tohjima, Y., Wakita, H. (1993). Remote detection of HCl and SO2 in volcanic gas from Unzen Volcano, Japan. *Geophysical Research Letters*, 20(13), 1355–1358.

Naboko, S.I. (1959). Volcanic exhalations and products of their reactions as exemplified by Kamchatka-Kuriles volcanoes. *Bulletin Volcanologique*, 20(1), 121–136.

Naughton, J.J., Derby, J.V., Glover, R.B. (1969). Infrared measurements on volcanic gas and fume: Kilauea eruption, 1968. *Journal of Geophysical Research (1896–1977)*, 74(12), 3273–3277.

Neuberg, J., Luckett, R., Ripepe, M., Braun, T. (1994). Highlights from a seismic broadband array on Stromboli Volcano. *Geophysical Research Letters*, 21, 749–752.

Nordlie, B.E. (1971). The composition of the magmatic gas of Kilauea and its behavior in the near surface environment. *American Journal of Science*, 271(5), 417–463.

Okubo, P.G., Nakata, J.S., Koyanagi, R.Y. (2014). The evolution of seismic monitoring systems at the Hawaiian Volcano Observatory. *Characteristics of Hawaiian Volcanoes*. doi.org/10.3133/pp18012.

Olivieri, M. and Clinton, J. (2012). An almost fair comparison between Earthworm and SeisComp3. *Seismological Research Letters*, 83(4), 720–727.

Omori, F. (1916). The Sakurajima eruptions and earthquakes, IV. Results of the levelling surveys and the Kagoshima Bay soundings made after the Sakurajima eruption of 1914 (in Japanese). *Bull. Imp. Earthq. Inv. Comm.*, 8, 322–351.

Oppenheimer, C., Francis, P., Burton, M., Maciejewski, A.J.H., Boardman, L. (1998). Remote measurement of volcanic gases by Fourier transform infrared spectroscopy. *Applied Physics B*, 67(4), 505–515.

Palmieri, L. (1874). Il sismografo portatile. *Atti Dell'Accademia Pontaniana Di Napoli*, 123–130.

Peltier, A., Staudacher, T., Catherine, P., Ricard, L.-P., Kowalski, P., Bachèlery, P. (2006). Subtle precursors of volcanic eruptions at Piton de la Fournaise detected by extensometers. *Geophysical Research Letters*, 33(6). doi.org/10.1029/2005GL025495.

Pingue, F., Troise, C., Luca, G., De Grassi, V., Scarpa, R. (1998). Geodetic monitoring of Mt. Vesuvius Volcano, Italy, based on EDM and GPS surveys. *Journal of Volcanology and Geothermal Research*, 82(1/4), 151–160.

Pingue, F., Berrino, G., Capuano, P., Obrizzo, F., Natale, G., De Esposito, T. et al. (2000). Ground deformation and gravimetric monitoring at Somma-Vesuvius and in the Campanian volcanic area (Italy). *Physics and Chemistry of the Earth, Part A: Solid Earth and Geodesy*, 25(9/11), 747–754.

Pingue, F., Bottiglieri, M., Godano, C., Obrizzo, F., Tammaro, U., Esposito, T., Serio, C. (2013). Spatial and temporal distribution of vertical ground movements at Mt. Vesuvius in the period 1973–2009. *Annals of Geophysics*, 56(4). doi.org/10.4401/ag-6457.

Powell, W.B. and Pheifer, D. (2000). The electrolytic tilt sensor. *Sensors*, 39(5), 120.

Pozzi, J.P., Mouël, J.L.L., Rossignol, J.C., Zlotnicki, J. (1979). Magnetic observations made on La Soufriere Volcano (guadeloupe) during the 1976–1977 crisis. *Journal of Volcanology and Geothermal Research*, 5(3), 217–237.

Richter, C.F. (1935). An instrumental earthquake magnitude scale. *Bulletin of the Seismological Society of America*, 25(1), 1–32.

Roche, O., Druitt, T.H., Merle, O. (2000). Experimental study of caldera formation. *Journal of Geophysical Research: Solid Earth*, 105(B1), 395–416.

Rose Jr., W.I. and Stoiber, R.E. (1969). The 1966 eruption of Izalco Volcano, El Salvador. *Journal of Geophysical Research (1896–1977)*, 74(12), 3119–3130.

Saleh, B. (1986). Développement d'une nouvelle instrumentation pour les mesures de déformations – Applications au génie civil. PhD Thesis, Université Pierre et Marie Curie Paris VI, Paris.

Sato, M. and McGee, K. (1981). Continuous monitoring of hydrogen on the south flank of Mount St. Helens. In *The 1980 Eruptions of Mount St. Helens*, Lipman, P.W., Mullioneaux, D.R. (eds). USGS Professional Paper, Washington.

Sheets, P. (2015). Volcanoes, ancient people, and their societies. In *The Encyclopedia of Volcanoes*, 2nd edition, Sigurdsson, H., Houghton, B., McNutt, S., Rymer, H., Stix, J. (eds). Academic Press, Amsterdam.

Shepherd, E.S. (1925). The analysis of gases obtained from volcanoes and from rocks. *Journal of Geology*, 33(3), 289–370.

Shimozuru, D. (1983). Volcano surveillance and prediction of eruptions in Japan. In *Forecasting Volcanic Events*, Tazieff, H., Sabroux, J.C. (eds). Elsevier, Amsterdam.

Shinohara, H. (2005). A new technique to estimate volcanic gas composition: Plume measurements with a portable multi-sensor system. *Journal of Volcanology and Geothermal Research*, 143(4), 319–333.

Siebert, L., Simkin, T., Kimberly, P. (2010). *Volcanoes of the World*, 3rd edition. University of California Press, Berkeley.

Sigurdsson, H. and Carey, S. (1986). Volcanic disasters in Latin America and the 13th November 1985 eruption of Nevado del Ruiz volcano in Colombia. *Disasters*, 10(3), 205–216.

Sigurdsson, H., Houghton, B., McNutt, B., Rymer, B., Stix, J. (eds) (2015). *The Encyclopedia of Volcanoes*, 2nd edition. Academic Press, Amsterdam.

Sindrason, S. and Olafsson, H. (1978). A magnetoresistor geotiltmeter for monitoring ground movement. Nordic Volcanological Institute [Online]. Available at: https://nordvulk.hi.is/sites/nordvulk.hi.is/files/NVI_Reports_pdf/nvi_report_7806_low_text.pdf.

Stoiber, R.E. and Malone, G.B. (1975). Sulfur dioxide emissions at the crater of Kīlauea, at Mauna Ulu, and at Sulfur Banks. *Eos, Transactions, American Geophysical Union*, 56(6), 461.

Sutton, A.J. and Elias, T. (2014). One hundred volatile years of volcanic gas studies at the Hawaiian Volcano Observatory. In *Characteristics of Hawaiian Volcanoes*, Poland, M.P., Takahashi, T.J., Landowski, C.M. (eds). doi.org/10.3133/pp18017.

Sutton, A.J., Elias, T., Kauahikaua, J. (2003). Lava-effusion rates for the Pu'u 'Ō'ō-Küpaianaha eruption derived from SO_2 emissions and very low frequency (VLF) measurements [Online]. Available at: https://pubs.usgs.gov/pp/pp1676/pp1676_08.pdf.

Swanson, D.A., Rose, T.R., Fiske, R.S., McGeehin, J.P. (2012). Keanakāko'i Tephra produced by 300 years of explosive eruptions following collapse of Kīlauea's caldera in about 1500 CE. *Journal of Volcanology and Geothermal Research*, 215–216, 8–25.

Tazieff, H. (1977). An exceptional eruption: Mt. Niragongo, Jan. 10th 1977. *Bulletin Volcanologique*, 40, 189–200.

Thomas, R.J., Krehbiel, P.R., Rison, W., Edens, H.E., Aulich, G.D., Winn, W.P. et al. (2007). Electrical activity during the 2006 Mount St. Augustine volcanic eruptions. *Science*, 315(5815), 1097.

Thorarinsson, S., Sigurdur Einarsson, T.S., Kristmannsdóttir, H., Óskarsson, N. (1973). The eruption on Heimaey, Iceland. *Nature*, 241(5389), 372–375.

Tilling, R.I. (1989). Volcanic hazards and their mitigation: Progress and problems. *Reviews of Geophysics*, 27(2), 237–269.

Tilling, R.I., Kauahikaua, J.P., Brantley, S.R., Neal, C.A. (2014). The Hawaiian Volcano Observatory – A natural laboratory for studying basaltic volcanism [Online]. Available at: https://pubs.usgs.gov/pp/1801/downloads/pp1801_Chap1_Tilling.pdf.

Tryggvason, E. (1982). The NVI magnetoresistor tiltmeter results of observations 1977–1981 [Online]. Available at: http://nordvulk.hi.is/sites/nordvulk.hi.is/files/NVI_Reports_pdf/nvi_report_8203_low_text.pdf.

Uyeda, S. (1961). An interpretation of the transient geomagnetic variations accompanying the volcanic activities at Volcano Mihara, Oshima Island, Japan. *Bulletin of the Earthquake Research Institute, University of Tokyo*, 39, 579–591.

Van Camp, M., de Viron, O., Watlet, A., Meurers, B., Francis, O., Caudron, C. (2017). Geophysics from terrestrial time-variable gravity measurements. *Reviews of Geophysics*, 55(4), 938–992.

Vergniolle, S., Boichu, M., Caplan-Auerbach, J. (2004). Acoustic measurements of the 1999 basaltic eruption of Shishaldin volcano, Alaska 1: Origin of Strombolian activity. *Journal of Volcanology and Geothermal Research*, 137, 135–151.

Villari, L. (1983). Volcano surveillance and volcanic hazard assessment in the Etnean area. In *Forecasting Volcanic Events*, Tazieff, H., Sabroux, J.C. (eds). Elsevier, Amsterdam.

Voight, B. (1990). The 1985 Nevado del Ruiz volcano catastrophe: Anatomy and retrospection. *Journal of Volcanology and Geothermal Research*, 42(1/2), 151–188.

Voight, B., Hidayat, D., Sacks, S., Linde, A., Chardot, L., Clarke, A. et al. (2010). Unique strainmeter observations of Vulcanian explosions, Soufrière Hills Volcano, Montserrat, July 2003. *Geophysical Research Letters*, 37(19). doi.org/10.1029/2010GL042551.

Weibring, P., Edner, H., Svanberg, S., Cecchi, G., Pantani, L., Ferrara, R., Caltabiano, T. (1998). Monitoring of volcanic sulphur dioxide emissions using differential absorption lidar (DIAL), differential optical absorption spectroscopy (DOAS), and correlation spectroscopy (COSPEC). *Applied Physics B*, 67(4), 419–426.

Westphal, J.A., Carr, M.A., Miller, W.F., Dzurisin, D. (1983). Expendable bubble tiltmeter for geophysical monitoring. *Review of Scientific Instruments*, 54(4), 415–418.

Williams, S.N. (1990). Nevado del Ruiz volcano, Colombia: An example of the state-of-the-art of volcanology four years after the tragic November 13, 1985 eruption. *Journal of Volcanology and Geothermal Research*, 41(1), 1–5.

Wilson, R.M. (1927). Surveys around Kilauea. *The Volcano Letter*, 128, 1.

Wilson, R.M. (1935). Ground surface movements at Kilauea Volcano, Hawaii. *University of Hawaii Research Publication*, 10.

Wood, H.O. (1913). Hawaiian Volcano Observatory. *Seismological Society of America Bulletin*, 3(1), 14–19.

Yamashita, K.M. (1981). Dry tilt: A ground deformation monitor as applied to the active volcanoes of Hawaii. Open-File Report. doi.org/10.3133/ofr81523.

Yamazaki, K.M., Teraishi, M., Ishihara, K., Komatsu, S., Kato, K. (2013). Subtle changes in strain prior to sub-Plinian eruptions recorded by vault-housed extensometers during the 2011 activity at Shinmoe-dake, Kirishima volcano, Japan. *Earth, Planets and Space*, 65(12), 1491–1499.

Yokoyama, I. (1986). Crustal deformation caused by the 1914 eruption of Sakurajima volcano, Japan and its secular changes. *Journal of Volcanology and Geothermal Research*, 30(3), 283–304.

Yukutake, Y., Yoshino, T., Utada, H., Shimomura, T., Kimoto, E. (1983). Changes in the apparent electrical resistivity of Oshima volcano observed during a period of highly elevated tectonic activity. *Earthquake Prediction Research*, 2(83/96).

Yukutake, Y., Yoshino, T., Utada, H., Watanabe, H., Hamano, Y., Sasai, Y., Shimomura, T. (1987). Changes in the electrical resistivity of the central cone, Miharayama, of Izu-Oshima volcano, associated with its eruption in November 1986. *Proceedings of the Japanese Academy, Series B*, 63, 55–58.

List of Authors

Patrick BACHÈLERY
Laboratoire Magmas et Volcans
CNRS, IRD, OPGC
Université Clermont Auvergne
Clermont-Ferrand
France

Philipson BANI
Laboratoire Magmas et Volcans
CNRS, IRD, OPGC
Université Clermont Auvergne
Clermont-Ferrand
France

Sylvain CHARBONNIER
School of Geosciences
University of South Florida
Tampa
United States

Oryaëlle CHEVREL
Laboratoire Magmas et Volcans
CNRS, IRD, OPGC
Université Clermont Auvergne
Clermont-Ferrand
France

Franck DONNADIEU
Laboratoire Magmas et Volcans
CNRS, IRD, OPGC
Université Clermont Auvergne
Clermont-Ferrand
France

Julia EYCHENNE
Laboratoire Magmas et Volcans
CNRS, IRD, OPGC
Université Clermont Auvergne
Clermont-Ferrand
France

Pierre-Jean GAUTHIER
Laboratoire Magmas et Volcans
CNRS, IRD, OPGC
Université Clermont Auvergne
Clermont-Ferrand
France

Mathieu GOUHIER
Laboratoire Magmas et Volcans
CNRS, IRD, OPGC
Université Clermont Auvergne
Clermont-Ferrand
France

Claude JAUPART
Institut de physique du globe
de Paris
Université de Paris
Académie des sciences
France

David JESSOP
Laboratoire Magmas et Volcans
CNRS, IRD, OPGC
Université Clermont Auvergne
Clermont-Ferrand
and
OVSG
Gourbeyre
Guadeloupe Institut de physique
du globe de Paris
Université de Paris
Paris
France

Karim KELFOUN
Laboratoire Magmas et Volcans
CNRS, IRD, OPGC
Université Clermont Auvergne
Clermont-Ferrand
France

Jean-François LÉNAT
Laboratoire Magmas et Volcans
CNRS, IRD, OPGC
Université Clermont Auvergne
Clermont-Ferrand
France

Séverine MOUNE
Laboratoire Magmas et Volcans
CNRS, IRD, OPGC
Université Clermont Auvergne
Clermont-Ferrand
and
OVSG
Gourbeyre
Guadeloupe Institut de physique
du globe de Paris
Université de Paris
Paris
France

Raphaël PARIS
Laboratoire Magmas et Volcans
CNRS, IRD, OPGC
Université Clermont Auvergne
Clermont-Ferrand
France

Olivier ROCHE
Laboratoire Magmas et Volcans
CNRS, IRD, OPGC
Université Clermont Auvergne
Clermont-Ferrand
France

Jean-Claude THOURET
Laboratoire Magmas et Volcans
CNRS, IRD, OPGC
Université Clermont Auvergne
Clermont-Ferrand
France

Index

A, B, C

D, E, F

G

H

I, J, K

Printed and bound by CPI Group (UK) Ltd, Croydon, CR0 4YY

19/04/2023

03212019-0001